The Sassafras Science Adventures

Volume 8: Physics

Johnny Congo & Paige Hudson

The Sassafras Science Adventures
Volume 8: Physics

First Edition 2025

Email: support@elementalscience.com

978-1-953490-29-2
Cover Design by Paige Hudson & Eunike Nugroho
Illustrations by Eunike Nugroho (be.net/inikeke)

Printed In USA For World Wide Distribution

For more copies write to:
Elemental Science
PO Box 4583
Maryville, TN 37802
support@elementalscience.com

Dedication

To the one who gave us the whole idea for this series. We hope we gave the Sassafras Twins the story you envisioned when you originally sparked the concept over dinner with friends.

Make the Most of Your Journey with the Sassafras Twins!

Add our activity guide, logbook, or lapbooking guide to create a full science curriculum for your students!

The Sassafras Guide to Physics includes chapter summaries and an array of options that coordinate with the individual chapters of this novel. This guide provides ideas for experiments, notebooking, vocabulary, memory work, and additional activities to enhance what your students are learning about the principles of physics!

The Official Sassafras SCIDAT Logbook: Physics Edition partners with the activity guide to help your student document their journey throughout this novel. The logbook includes their own SCIDAT log pages as well as lab report sheets and an physics glossary.

Lapbooking through Physics with the Sassafras Twins provides a gentle option for enhancing what your students are learning about the principles of physics through this novel. The guide contains a reading plan, templates and pictures to create a beautiful lapbook on physics, vocabulary, and coordinated scientific demonstrations!

Visit ElementalScience.com to learn more!

Table of Contents

AUTHORS' NOTE

The novel you hold in your hands is both a work of fiction and a teaching tool, which creates the need for a delicate balance.

We have taken fictional liberties when it comes to the story line—to our knowledge, no one is traveling around the world on invisible zip lines. But, when it comes to the information shared by our local experts, we have done our best to accurately reflect the current scientific knowledge at the time of writing this book.

We have drawn from our personal experiences and education, as well as the following resources:

- *Basher Science Physics: Why Matter Matters!*
- *Basher Science Engineering: The Riveting World of Buildings and Machines*
- *DK Children's Encyclopedia*
- *DK Encyclopedia of Science*
- *Kingfisher Science Encyclopedia*
- *The Usborne Children's Encyclopedia*
- *Usborne Science Encyclopedia*

In addition to these books, we have drawn on the information found on many different websites.

Please keep in mind that science is ever-changing as our technology advances and our understanding deepens. If you find an issue in this novel, please let us know and take the opportunity to discuss these new developments with your students.

The Sassafras Guide to the Characters

Cecil's Neighborhood (Chapter 1)

- ☆ **Mr. Ernest Womberfraggle§** – He's a middle school science teacher who once had students named Cecil, Summer, Thaddeus, Paul, and Yang Bo.
- ☆ **Mrs. Wilma Womberfraggle§** – She's the local librarian.
- ☆ **Tracey Sassafras§** – The female twin of the almost famous duo known as the Sassafras twins who models, solves riddles, and operates cameras. She is also known as Blaisey, Fish Hook, and Tracey the Plucky.
- ☆ **Blaine Sassafras§** – The older and handsomer, in his own mind, twin of the almost famous duo known as the Sassafras twins. He is also known as Train, Rowboat, Blaine the Handsome, and the Little Brother.
- ☆ **Cecil Sassafras§** – The inventor and scientist responsible for the twins' summer of science. He is also the Sassafras twins' uncle who will probably never get their names right.
- ☆ **President Lincoln§** – The brilliant, but quiet prairie dog who serves as Uncle Cecil's right-hand paw. He is the legendary mammal also known as The Prez and Linc Dog.
- ☆ **Thaddeus Wazeezy§** – Also known as The Man With No Eyebrows and the Dark Cape. His plans for revenge are about to culminate in the most forgettable way.

Papua New Guinea (Chapters 2-3)

- ☆ **Summer Beach§** – She is a most unique, sandwich-loving scientist! She has brought her infectious energy to every leg of the twins' journey so far. She's Cecil's schoolmate turned best friend and a *pro re*

§ These characters appear throughout the novel. We have chosen to share about them in the chapter where they first appear.

* These characters are only mentioned in the text but are not a part of this leg of the journey.

nata agent for the Triple S (Swiss Secret Service).

- ✯ **Paul Sims**[§] – He is the museum curator at the National Air and Space Museum. He is also a friend and schoolmate of Cecil's and Summer's who is hiding quite a bit.
- ✯ **Mrs. Pascapali (paz-kah-pah-LEE)**[§] – She is Uncle Cecil's neighbor who lives at 1106 North Pecan Street.
- ✯ **Susa Dasini** – She is the chief scientist and engineer at Centripetal Village.
- ✯ **Polly Glot** – She is Susa's friend and in charge of marketing at Centripetal Village.
- ✯ **Bavo Becker** – He is the dour potential investor in Centripetal Village.
- ✯ **Alwin Aldrich** – He is a friend of Mr. Becker and another potential investor.

Camp Zip Fire in Montana (Chapters 4-5)

- ✯ **Hattie Cates** – One of the twin's local experts for their time in Montana. She is also a regular Camp Zip Fire counselor for the Appaloosa tribe.
- ✯ **Joanna, Janna, Marci, and DeeDee** – Tracey's camp buddies from their years at Camp Zip Fire.
- ✯ **Chaz Throckmorton** – One of the twin's local experts for their time in Montana. He is also a regular Camp Zip Fire counselor for the Mustang tribe.
- ✯ **Jamal, Jarod, Matt, and Davy** – Blaine's camp buddies from their years at Camp Zip Fire.
- ✯ **Grizz** – The bearded and burly Camp Zip Fire staffer, who does anything and everything required to keep the campground up and running.
- ✯ **Timmy and Gretta Brown** – The brother and sister who join the Appaloosa and Mustang tribes for that summer.

Brazil (Chapters 6-7)

- **Augustus McShane** – A science-loving special agent who serves as the twins' local expert for their time near Iguazu.
- **Adelina Ferreira** – A tour guide with Ferreira Tours. She leads the tour of Iguazu Falls and Itaipu Dam.
- **Ox** – He is the helicopter pilot provided by Colt and Barlow.
- **Colt Belford** – He is one half of the tourist duo who aren't what they seem.
- **Barlow Hennessy** – He is the other half of the tourist duo who aren't what they seem.

Greenland (Chapters 8-9)

- **Ari Jensen** – He is a cameraman who serves as the twins' local expert for their leg in Nuuk.
- **Rozing Kleist** – She is the talent manager for Stependous.
- **Hans Lyberth** – He is the mayor of Nuuk, Greenland.
- **Justus** – He is the frontman of Stependous.
- **Faro** – He is known as the heartthrob of Stependous.
- **Anse** – He is known as the bad boy of Stependous.
- **Pilu** – He is known as the shy one of Stependous.

Spain (Chapters 10-11)

- **Esperanza Tambor** – She is an orphan who serves as their local expert for the twins' leg in Barcelona.
- **Alejandra and Maya** – Two of the girls who live with Esperanza.
- **Valentina** – A taller girl, who serves as the leader of sorts for the ragtag group of orphans in Spain known as *Los Granujos.*
- **Maria Aldana** – She works at the *Palau de la Musica Catalana* and also helps the group of ragamuffins.
- **Maestro Martinez** – He is the conductor at the *Palau de la Musica Catalana.*

South Africa (Chapters 12-13)

- ★ **Lubanzi LeRoux** – He is the radio host for Cape Town the Whole Night Through and the twins' local expert in Cape Town.
- ★ **John Z** – The audio mixer for the Cape Town the Whole Night Through radio show.
- ★ **Chester and Lenny** – Owners of Chester and Lenny's Good Morning Products, one of the sponsors of the Cape Town the Whole Night Through radio show.
- ★ **Dakari the Truckdriver** – A faithful listener and frequent caller for the Cape Town the Whole Night Through radio show.
- ★ **Lindiwe** – A cat lady who often calls into the Cape Town the Whole Night Through radio show.
- ★ **Junior** – A rugby player who often calls into the Cape Town the Whole Night Through radio show.
- ★ **Bandile Botha** – A former rival DJ to the Cape Town the Whole Night Through radio show.
- ★ **The Candy Line** – A group of night shift line workers at the Cape Candies food processing plant who often call into the Cape Town the Whole Night Through radio show.
- ★ **Kati the Go-Getter** – A lady who often calls into the Cape Town the Whole Night Through radio show.

CAMBODIA (CHAPTERS 14-15)

- ★ **Jaya (J.T.) Tang** – He loves science and math, but doesn't share the same passion with his brothers. He serves as the twins' local expert while they are in Siem Reap.
- ★ **Kosal** – The foreman of the work crew at Angkor Wat.
- ★ **Captain Marolf**§ – He is the head of the Triple S. He also appeared in the twins' earth science and astronomy legs.
- ★ **Evan DeBlose**§ – The lead Triple S agent and earth science local expert, who took down the rogue Agent Adrianne Archer and the nefarious Yuroslav Bogdanovich.
- ★ **Q-Tip**§ – The Triple S's expert in technologizing and good friend of Agent DeBlose.

Mexico (Chapters 16-17)

- ✯ **Verano Playa** – The mysterious local expert during the twins' visit to Mexico. They have met her multiple times in their adventures, but they usually find her in a much cooler place.
- ✯ **El Asistente** – This mythical Arctic ground squirrel lends a paw and hangs out in robotics forums.
- ✯ **Matteo Perez** – The wheelchair-using friend of Summer. He has designed *El Fausto*.
- ✯ **Solana** – Matteo's mother.
- ✯ **Choncho Caldas** – Matteo's competitor in the *Lucha Libre* contest. He has designed *El Caliente Picante*.
- ✯ **Gordo Gomez** – Matteo's friend and competitor in the *Lucha Libre* contest. He has designed *El Burrito*.
- ✯ **Gonzalo Viejo** – The referee for the *Lucha Libre* contest.
- ✯ **Kevin Maye** – Matteo's competitor in the *Lucha Libre* contest. He has designed *El Texican*.
- ✯ **REESE** – The robotic creation of President Lincoln and Ulysses S. Grant. His name stands for Robotic Exploration, Entertainment, and Scientific Enhancement.

Pecan Street (Chapter 18)

- ✯ **Yang Bo** – He is the astronaut who served as the twins' local expert on the International Space Station. He is also a former classmate of Uncle Cecil and Summer from middle school.

Volume 8

Physics

CHAPTER 1: A TRIP TO THE LIBRARY

Poetic Presentation

A match made in heaven. Anybody could see they belonged together. Like bacon and eggs. Like a tire swing and a tree. Like a novel and an easy chair. Like eyebrows and a forehead. These two belonged together.

With enraptured smiles on their faces, the swooning couple walked hand in hand down the quaint neighborhood sidewalk. A fresh morning breeze and sprawling rays cast out by a new sunrise accompanied them as they took unrushed steps, fulfilled and content in each other's company. Their clasped hands swung back and forth as they walked, matching the rhythm of their smiles. What a lovely thing to start the day off like this, walking together. They turned from the picturesque street of homes and yards, continuing down the sidewalk along a slightly bigger road.

Now, instead of sauntering by cute little places of residence, they were passing hospitable places of business: The Left-Handed Turtle Neighborhood Market, The Right-Handed Pelican Pizza Parlor, The Ambidextrous Octopus Bowling Alley. The two lovebirds were not in any hurry, but eventually they needed to get to where they were going. Sadly, that was to two separate locations.

She would stop after the next block, and he would continue for three. Parting would be such sweet sorrow, but these two were used to it. She was the head librarian, and he was the sole junior high science teacher: Mr. and Mrs. Womberfraggle. Ernest and Wilma. They'd moved here as newlyweds and trekked this way every day for almost forty-five years. Over the years they'd realized just as they were a perfect match for each other, they were also a perfect match for this quaint, little town.

The years had been kind to the Womberfraggle's and their little city. The people and the town were older and maybe required a little more maintenance than before, but they were also still healthy and full of cheerful optimism. The duo floated affectionately over their remaining distance together. Ernest then looked at his wife, lifted the hand he was still grasping, and kissed it.

"Goodbye, my love," he said with a smile. "Have a wonderful day."

Wilma's countenance reflected his. "You too," she said as she let her hand drop from his.

She then turned to look both ways before crossing the crosswalk that would take her to the library. After arriving on the other side safely, she turned to give one last loving wave to her husband. Ernest reciprocated and then walked three more blocks to the junior high school.

Wilma reached into her purse to grab her ring of keys. She was always the one to open the library in the morning, so she didn't expect anyone. But as she approached the door, she saw them. One in a messy white lab coat and bunny house slippers. Two in T-shirts and sneakers. And another was a prairie dog.

"Wowie wee willikers and a wonderiffic morning to you, Mrs. Womberfraggle," the enthusiastic red-headed, happy-faced, lab-coat-wearing oldest member of the group said with a grin.

"Well, good morning, Cecil," Wilma returned the greeting. She then acknowledged the other three in the party. "I recognize President Lincoln, but who are these other two?"

"Why, this is my nephew and my niece, son and daughter of my older brother, Train and Blaisey Sassafras!" Cecil exclaimed.

"Train and Blaisey?" Mrs. Womberfraggle asked pleasantly with a clarifying tone.

Uncle Cecil smiled triumphantly. Tracey Sassafras nodded and smiled, a little embarrassed by her uncle.

Blaine nodded, smiled, and held out his hand to greet the librarian. "Blaine and Tracey," he stated their names correctly. "We are the twelve-year-old twins hanging out with Uncle Cecil this summer, learning all about science!"

Realization found its way to the librarian's expression as she shook the boy's hand and then his sister's. "Oh, yes, that's right. Your uncle mentioned that you two were coming for a visit. Well, how's it going so far?"

"Zip-tastic," Blaine answered exuberantly.

"Zip-tastic?" The elderly librarian needed clarification again.

"It's going very well," Tracey jumped in for her brother. "We have learned so much about zoology, anatomy, botany, earth science, geology, astronomy, and chemistry. It has all been so stupendous!"

"And we're hoping for more," Blaine added.

"That is wonderful to hear," Wilma Womberfraggle gleamed. "It is so nice to see youngsters so eager to learn."

The librarian looked from the children back to their uncle. "Your brother should be very proud of his son and daughter, as you should be proud of your nephew and niece."

"I slippity-slappity am!" Cecil responded in his normal-to-him, slap-dash style.

The white-haired librarian nodded. "Oh yes, Cecil. Now I remember. You wanted to use the library this morning for a presentation of some kind. Isn't that right?"

"It sure is, Mrs. Womberfraggle," Cecil beamed. "Welly welp, it's more President Lincoln's presentation. But yessity, yes, we are here early this morning, bright-eyed and bushy-tailed, for a scientific video presentation!"

Mrs. Womberfraggle nodded and smiled as she used the correct key to open the library door. Blaine, being a gentleman, held it open for her to enter first. Wilma took a couple of steps into the library's entryway. Then she turned back and looked at the twins with a chuckle. "We sure appreciate your uncle here at the library, but if he's going to do any kind of studying, researching, or presenting, we have to let him do it early in the morning before anyone else gets here. His volume gets pretty high when he's excitable."

The twins knew exactly what the librarian was talking about. Cecil Sassafras was definitely one-of-a-kind. Before visiting this summer, Blaine and Tracey hadn't known their pseudo-famous research scientist uncle well, but over the past seven-ish weeks, they had come to know him and all his antics very well.

During the previous school year, Blaine and Tracey had done well in every subject except one: science. They had failed science, which had set their concerned parents' plan into motion. Blaine and Tracey's mom and dad had sent their children down to Cecil's house for the entire summer to do nothing but science.

Initially, this decision devastated the twelve-year-olds. They had hoped to spend their summer at their favorite place in all the world: Camp Zip Fire, a summer camp where virtually all of their friends went. It had a smorgasbord of awesome activities and a staff of the coolest counselors. Instead, they thought they were going to spend a whole summer with no friends and boring learning with a nerdy scientist. Needless to say, they didn't have the greatest attitudes when their parents put them on the long-distance bus and shipped them to Cecil's place.

However, over the course of the summer, their attitudes toward both the subject of science and their uncle had changed drastically. Uncle Cecil, along with his prairie dog lab assistant, President Lincoln, had invented invisible zip lines. Zip lines that, when used along with custom carabiners, could zip a person anywhere on the globe at the speed of light. When Cecil and President Lincoln had first presented the fantastical concept to them, Blaine and Tracey thought it was some sort of ridiculous joke, but then they had tried it, and it worked! All summer they had donned helmets, cinched up harnesses, calibrated the three rings on the specially designed carabiners and had been zipping to locations all over the planet.

One of the carabiner rings was for longitude, one ring was for latitude, and one ring was to lock the carabiner shut. As soon as the carabiner snapped shut, it immediately connected to the specific invisible zip line that would take them to the set longitude and latitude coordinates. The twins would hang in mid-air a couple feet off the ground for approximately seven seconds, and then, whoosh! Off they would zip at the speed of light. The first location they had zipped off to had been Kenya, where they had plunged into the subject of zoology. The Sassafras twins hadn't looked back since!

Cecil's heart behind the invisible zip lines was to create an adventurous way for his young relatives to study science face to face so they would hopefully go from hating science to loving it. The plan had worked splendidly, and that's exactly what had happened over the course of the summer as the twelve-year-olds zipped back and forth, experiencing adventure and gaining knowledge in multiple subjects of science.

Back at the library, the lovely mustiness of shelved books met the twins' noses like a familiar friend. The dusty notes mixed with hints of smoke, cedar, and vanilla permeated their senses as the two children entered behind the two adults and the one prairie dog. Blaine and Tracey buzzed with excitement over their long-awaited library visit.

Rows and rows of books greeted them, but they also saw cozy-looking chairs and sofas. Against the back wall stood wooden tables and stools for group study and research. The walls held colorful signs and posters touting the joys and benefits of reading, along with displays of unique local artwork, some crafted by adults and some by children. There were bean bags, carpet squares, hidden nooks, cozy corners, classic desktop computers, and even a couple old microfiche machines. This place was amazing!

At the back of the house of books sat a small semicircle stage apt for giving presentations, and that is where Mrs. Womberfraggle was leading them now. The white-haired woman reached the spot and then gestured toward it with her hand.

"Here you go, Cecil," she said with a smile. "It's all yours. Have fun, but remember, this is a library."

"Dokie-oke. Mrs. Womberfraggle, thankiny thank you so much!" Cecil responded exuberantly and loudly, not picking up the librarian's gentle hint.

The kind woman shook her head, smile still intact. She turned to acknowledge the twins. "It was nice to meet you, Blaine and Tracey. I hope you continue to enjoy your summer."

"Oh, we will!" Blaine responded happily and with certainty. Then Tracey added, "Nice to meet you too, Mrs. Womberfraggle."

The matronly woman turned and began making her way through the maze of books, back toward her desk at the front of the library.

Blaine and Tracey turned toward the small stage, and when

they did, they saw that somehow Uncle Cecil and President Lincoln were already ready for their presentation. The twins' wild-haired, red-headed uncle was standing in his spot on the platform, ready to serve as the orator—something he had done for each and every one of President Lincoln's presentations.

President Lincoln perched atop a small control panel off to the side, ready to tap any needed buttons. Even after seven weeks of observing and interacting with the Prez, the Sassafras twins still weren't sure how the furry mammal contributed so much to their uncle's work. Was he really accomplishing everything their uncle was giving him credit for? The twins weren't sure.

A screen affixed to the back of the stage already displayed the first bright, fun, and colorful image of the coming presentation. The image showed the rodent with outstretched paws and an outstretched smile. Text arched over his head which read, *President Lincoln's Ever-so-brief Presentation on Chemistry*!

Just yesterday, the young science learners had returned to their uncle's house and neighborhood after zipping to eight different invisible zip-line locations to study different topics within the subject of chemistry with the help and guidance of brilliant and capable local experts. Now they would get to see a picture-filled presentation briefly summarizing what they had learned, evidently all organized and put together by President Lincoln.

Their uncle's basement lab, which usually hosted these presentations, was unavailable at the moment. The mammal tapped a button over at the control panel, and the next image came up on the screen: an image of the periodic table.

"The periodic table is a systematic arrangement of the elements, grouping those with similar properties together." Their uncle read the text written above the image in a semi-normal voice. While performing the duties of orator for these ever-so-brief presentations, Cecil channeled a deeper, more professional voice.

"The alkali metals are six very reactive metals. These metals are soft and shiny, but they change color as soon as they hit the air." Cecil read the text over the next image, and then he continued, "The alkaline earth metals are less reactive than the alkali metals, but they still easily combine with oxygen to form many of the compounds found in the earth's crust."

Over at the control panel, President Lincoln kept tapping a button. Every time he did, the image and text would change.

"The transition metals are a group of elements found in the middle of the periodic table. For the most part, these elements are useful, hard, and shiny. Some examples are iron, copper, gold, silver, and chromium." The images from the periodic table appeared. They were the symbols—the 1-, 2-, or 3-letter codes—that scientists use for the element Cecil was reading the information about.

"The lanthanides are a group of fifteen naturally occurring heavy but soft metals. In modern times, we use lanthanide elements in touchscreens, wind turbines, and batteries." An image of the list of lanthanide symbols came up as Cecil read this part. The letters and numbers reminded the twins of the time they had spent in Japan studying these elements with a ninja Sensi and his dojo full of pupils.

"The actinides are a group of fifteen heavy metals. Many of the actinides are radioactive, which means they easily break apart, releasing energy and potentially damaging particles." Blaine and Tracey exhaled in relief as their uncle read the information about actinides, remembering how they, along with their ninja friends, had helped to stop a nuclear power plant from exploding out on a dark island.

Next, their uncle read about a couple of elements the twins had learned about in Singapore. "Main group metals, also known as ordinary metals, base metals, or poor metals, can conduct electricity and heat well. They are also easy to shape."

Tracey blushed as she thought about the chance she had to be a runway model in a Singaporean fashion show. One that had showcased many of the main group metals, and also some metalloids, which the red-headed scientist announced next. "Metalloids, also known as semi-metals or semiconductors, look like metals and even have some metallic properties, but they are often too brittle and don't like to behave like metals typically do."

President Lincoln gave another tap of the button, and another image appeared from the periodic table. "Nonmetals can be gases, liquids, or solids, and they behave differently in chemistry. Five of them are essential to life," Cecil read in his refined baritone voice, following fastidiously for the next image. "The halogens are all nonmetals that react strongly. They easily 'steal' electrons from metals, especially the alkali metals, bonding together to form salts. The name halogen literally means 'salt giver.'"

"And last but not least," the research scientist read after the prairie dog tapped the button to reveal the last image of the presentation, "Noble gases."

The image reminded the twelve-year-olds of their escape by cover of night in a hot-air balloon in Morocco.

"The noble gases are inert, meaning they don't really react chemically or combine with other elements. Many of the noble gases float around in the atmosphere."

Blaine and Tracey sighed contentedly. They had worked so hard and learned so much. A sense of satisfaction filled them as they saw the elemental symbols and heard the information they had entered. They smiled at the memory of all the places they had invisibly zipped to and all the amazing people they had met along the way.

"This concludes *President Lincoln's Ever-so-brief Presentation about Chemistry*," Cecil Sassafras said with perfect posture in his refined orator's voice.

Foundational Physics

Switching back to his normal, spastic self, Cecil said, "Train and Blaisey, you did it! Isn't it wonderiffic, you did it! You two rickety rocked your way through the study of chemistry, and what splendimental images you sent in for all those elements using the archive application on your smartphones! Fantabulous job, you two!"

The Sassafras twins smiled sheepishly at all the over-the-top compliments coming from their uncle, fidgeting with the smartphones he'd just mentioned. Uncle Cecil and the Prez had invented the invisible zip lines. The man and the mammal had come up with this mode of study and had also had the idea to use and include these sleek phones.

The smartphones held the keys they needed to learn about science and travel on the invisible zip lines. The devices kept the twins connected to their science-loving uncle. They could reach him from any global location they found themselves. In real time, he could track them and they could call him. The smartphones also held a range of useful applications that helped them proceed through their summer of learning.

The LINLOC application, which stood for "Line Locations," gave the twelve-year-olds the specific longitude and latitude coordinates around the globe they would zip to. The app also shared the scientific topics they would study and the name of the local expert who would assist them. The SCIDAT application, which stood for "Scientific Data," was used to record the specifics of every scientific topic Blaine and Tracey learned. They would input the information using the app and then send it to their uncle, who could view the data on a large tracking screen in the basement lab he had at his residence. To help with the images required for the SCIDAT data, the phones contained three image-capturing assistants: the archive app, the microscope app, and the high-resolution cameras. During their study of chemistry, the twins had mainly used images they had found in the archive app to send in along with all their texted SCIDAT, but in previous adventures they had used mainly the cameras and occasionally the microscope app.

The twins' smartphones also came equipped with compass applications so they could keep their bearings straight when needed. Then, finally, the Prez had equipped their phones with the taser application. Unbelievably, this app could project jolts of electricity strictly used for protection in the most dangerous of situations. One would think that absolutely everyone would be thrilled about children learning science face to face, but sadly not everyone was.

Since their very first science-learning stop in Kenya, a silent villain had been stalking Blaine and Tracey all over the globe and even beyond. The Man With No Eyebrows was a sinister, scowling man with no hair upon his brow. For reasons unknown to the twins, he had been trying to stop them from learning science every step of the way. Not only could he somehow travel on the invisible zip lines like they did, but he also wore a Dark Cape suit, which he had stolen from a magician named Phil Earp. The suit gave him the ability to operate unseen. Blaine and Tracey had been more than a little afraid of the man at first, but now they just wanted to figure out who he was and understand why in the world he was trying to stop them from learning science.

Back in the library, the Sassafras twins shook away any thoughts of the Man With No Eyebrows. They grabbed their phones and their backpacks and followed their uncle and his lab assistant. With the study of chemistry finished, Blaine and Tracey were eager to see what was next!

Why in the world was he trying to stop them from learning science? To find the answer to that question, all he had to do was reach up and rub the fleshy padding of his index finger along the anger-wrinkled skin topping his eyeballs.

When he was born, he had been called Thaddeus Wazeezy. A good name. A family name. But those twins called him, "The Man With No Eyebrows." Why? Because he was a full-grown male with no hair between his nose and his skull. His eyebrows had been stolen long ago.

And who was the eyebrow burglar? None other than Cecil Sassafras. The ridiculous uncle of those twins.

Back in junior high school, Cecil had caused a Bunsen burner to explode. The explosion had taken Thaddeus' eyebrows away forever and had replaced them with shame. Over the years, the shame had festered into hate and bitterness. Now, Thaddeus' sole goal in life was to exact revenge on Cecil for robbing him of the joy of an eyebrow-filled life.

What was the best way to exact that revenge? The answer to

this question had come as a summer full of frustration and failed attempts. All of Cecil Sassafras's work as an eccentric scientist crescendoed into his grandest of grand inventions: invisible zip-line travel. And his greatest of great ideas: use the invisible zip lines to help his beloved niece and nephew learn science. Thaddeus had thought if he could ruin those twins' summer and stop them from learning science, this would derail Cecil, crushing the man's hopes and dreams. Thus, pouring out a full measure of vengeance for the eyebrows he had lost.

However, those twins had proven to be too resilient and too resourceful to stop. Even when Thaddeus had secured the use of an immaculate underground science lab, space-age technology, satellite capabilities, a horde of scientists working for him, and an army of Dark Cape suits, those twins had escaped. Yes, it had been a frustrating summer full of attempted revenge.

Thaddeus Wazeezy sat at his lonesome residence two doors down from where Cecil lived, back in the town where they had both attended junior high school. He hadn't been able to stop the twins, and he no longer had an immaculate lab, notable technology, satellites, scientists, or any kind of army. He had failed.

Or had he?

He still had his own Dark Cape suit. But even more, he had the Forget-O-Nator. This lone item survived the destruction of his Siberian underground lab. A mobile-home-sized machine designed to wipe away memories. And it currently sat in his backyard. Usher the target subjects inside the machine, lock the machine, turn the one and only knob to full capacity, and within moments, it would wipe all memories clean.

Yes, he had failed to stop the twins thus far, but once they had been Forget-O-Nated, his failure would cease. Everything they had learned so far this summer would be gone.

A wickedly happy wrinkle shoved its way into the middle of all the angry wrinkles on his hairless brow.

Evidently, this was the height of fashion in men's swimwear during the 1920s. The red-headed scientist had quickly stopped by the lavatory on his way out and had popped back out looking like this. The man was wearing a blue and white horizontally striped 1920s style swimsuit that covered him up from his shoulders to his knees. He accessorized with fins, goggles, a snorkel, and gobs of sunscreen. This was the first time the twins had ever seen their uncle without his pink bunny house slippers. However, he had an inflated pink bunny floatie secured around his waist.

The Sassafras twins wanted to be embarrassed, but they appreciated the novelty of the swimwear. Mrs. Womberfraggle waved goodbye with a smile as they exited the library. She didn't bat an eye at their uncle's outfit. This sort of thing must've appeared normal, at least for Cecil.

"Goodbye, Mrs. Womberfraggle," the twins said in unison. "Thanks for letting us use the library."

The elderly woman smiled. The group of four reciprocated, and from the treasure trove of books back out to beams of sunshine they went.

Blaine, Tracey, and President Lincoln looked just as they had when they had entered the library. Cecil looked altogether different.

"Toodapoo eego! Nek-uph: fizzegs!" Cecil mumbled with excitement, albeit inaudibly because of the snorkel in his mouth.

"Huh?" Blaine asked.

"Toodapoo eego! Nek-uph: fizzegs!" Cecil exuberantly shouted one more time, like what he had spoken was clear as day.

"What?" Tracey wanted to know. Obviously, the twins still didn't understand.

Slightly embarrassed for his human friend, President Lincoln, with a smile on his face, scurried up and pulled the snorkel out of the scientist's mouth. One more time, Cecil made his gleeful announcement, "To the pool we go! Next up, physics!"

The Sassafras twins looked at each other with wide eyes full of excitement and anticipation. They weren't sure what going to the pool had to do with anything, but they knew what, "Next up, physics!" meant. It meant they were about to hop on the invisible zip lines again to start an adventure to learn physics!

"Linc-Dawg and I thought the neighborhood pool would be the perfect place to transition from your successful study of chemistry to your upcoming study of physics!" Cecil explained. "Swimming pools are full of a compound known as H2O."

"Water!" the twins exclaimed, knowing this was the chemical symbol for water.

"Persactly!" confirmed their uncle. "To the pool we go, where we will jump from chemistry and dive into physics. H-two-Oh man, am I excited about this onboarding into a new subject!"

After several blocks of the flip-flopping and slip-slapping sound of Uncle Cecil's swimming fins hitting the sidewalk pavement, the quartet found themselves at the neighborhood pool. President Lincoln and the twins stood poolside, watching Cecil, who was standing on the diving board, still decked out in all his glorious swimwear.

"Nphins fed lah mph mosun stay—" Cecil said from his springy perch, but stopped when he saw all three of his companions pantomiming for him to take the snorkel out of his mouth.

The red-head laughed heartily and removed the piece of equipment from his mouth. "Well, slippery slapping snorkelarious! I keep forgetting that little accouterment is in there!" Cecil made sure the snorkel was out of his mouth, but it stayed attached to his goggles.

"Back to introducing physics!" The happy-go-lucky scientist started again. "Newton's third law of motion states that for every action, there is an equal and opposite reaction."

"For instance," Cecil continued, snorkel free, "On a diving board, a diver will put energy into the board by jumping on it. The board will then transfer that energy back into the diver, propelling them up and into the pool."

The elder Sassafras took a deep breath, smiled, and then said, "I will attempt to show this phenomenon now."

He took a few steps back to make sure he had the full length of the board for his approach. He leaned forward like he was about to go for it, but then paused as a slight frown made its way directly into the middle of his smile. "I sure dilly dally wish Summer was here with us to see this and to help me introduce physics to you two."

Little frowns found their way to Blaine's and Tracey's faces as well. They wished Summer Beach, their all-time favorite local expert, was here now too. In their minds, this is exactly where she should be. But she wasn't. Frowns were a rarity for Uncle Cecil. It evaporated quickly, replaced by his normal smile.

"Bowacunga!" The red-head shouted as he shot out toward the end of the diving board.

Amazingly, even with all his bulky swimming accessories, Cecil made it to the board's end with graceful agility. He jumped up, then came back down, firmly planting his flippers onto the surface of the diving board. The board returned his energy, launching him into the air, where he hung suspended for a second over the diamond-like, sparkling surface of the H_2O.

CHAPTER 2: PROPELLED TO PAPUA NEW GUINEA

Force Falls

Splash!

"Oh, c'mon," he scolded himself internally for not seeing the puddle that lay directly in front of the mailbox. Thaddeus pulled his shoe out of the dirty, shallow water and shook it with disgust. He gathered himself, opened up the street side mailbox at 1104 North Pecan Street, and put the invitation inside.

The very official looking envelope held an invitation to a special event. This invitation would set his plan into motion. Thaddeus wasn't just going to wipe away the memories of those twins—he was going to wipe away the memories of all the people involved. If the plan worked, and he was sure it would, Cecil Sassafras would check his mailbox and find this invitation inviting him to a junior high class reunion. He knew Cecil would be thrilled and would most certainly attend. Thaddeaus would also make sure both Summer Beach and Paul Sims received invitations as well. The reunion was set to occur right after those twins returned from the zip-learning adventure. He was sure they would be present at the event also. Even President Lincoln, Ulysses S. Grant, and REESE the robot would be there.

In the meantime, Thaddeus planned to discreetly move the large Forget-O-Nator to the junior high campus. He was going to work on a way to lock the machine and then start it from the inside. On the day of the reunion, when the despicable posse arrived at the school, they would be led to enter a certain classroom. However, instead of being a classroom, they would actually be walking into his memory-wiping machine.

He would greet them, revealing his identity to them. He would lay bare the guilt of Cecil Sassafras, reminding everyone how the red-head had caused a Bunsen burner to explode in his face, incinerating his eyebrows forever. He would then crank the machine to life, causing all of their memories to be erased. Cecil and his ingenious mind full of science would be blank. Summer's and Paul's memories would be zapped. The memories of those rodents would be toast. That robot's memory would be gone. And the twins. All the science they had learned this summer would be forget-o-nated into nothing.

And he himself, Thaddeus Wazeezy, the Man With No Eyebrows, with all his mistakes, shortcomings, and failures would have his memory wiped away into nothingness along with the others.

He shut Cecil's mailbox and sighed through a satisfied grimace, "What a glorious end this is going to be," he whispered resolutely. "For us all."

Splash!

The expensive flavored water gurgled as it poured into Summer Beach's glass from a bottle being held by Paul Sims. The museum curator filled her glass to the brim and then poured a glass for himself before sitting down to join her at the patio table.

"Beautiful day, isn't it?" Paul declared, drawing out the word beautiful.

Summer nodded and smiled politely.

The two were sitting at a table in the sunshine on Mrs. Pascapali's—the elderly widow who lived next door to Cecil—backyard patio. Yes, Summer smiled on the outside, but the smile didn't reach her eyes. Confusion more than clarity seemed to be what mostly was churning around internally. Yesterday she had been so hopeful and excited and joyful.

She and Ulysses S. Grant, her arctic ground squirrel lab assistant, had finished up the projects they'd been working on, along with completing their Triple-S, Swiss Secret Service, sanctioned directives. They had strapped on their harnesses and had ridden the invisible zip lines from Summer's Alaskan science lab to the front porch of 1104 North Pecan Street.

Summer had stood on the porch, bubbling over with excitement. She knew that a trapdoor was about to open up: a trapdoor that led to a slide, a slide that would lead to a reunion between her and Cecil Sassafras down in his basement lab.

"What will the reunion be like?" she'd wondered. The last time they'd been together had been at the Ambidextrous Octopus Bowling Alley, where they had shared a deeply meaningful moment under the sparkling light of a disco ball. She had loved Cecil since junior high school. But what about him? Was it love that she saw reciprocated in his eyes? He had started a stumbling sentence there in the bowling alley. Had he been attempting to confess his love for her? Summer didn't know.

The ringing of her smartphone had cut the beautiful moment short. It was a call she'd been compelled to answer, a call from Captain Marolf of the Swiss Secret Service. Answering the call had led to duties that necessitated her immediate departure—her departure from Cecil. But she had returned, more than ready to be reunited with him. She was more than ready to hear the completion of his sentence. More than ready to discover if this love she had long waited for was finally going to be revealed.

The trap door had opened, and she had squealed with uncontrollable joy all the way down the slide. But then, to her shock, the reunion had been nothing like what she'd imagined or hoped for. The basement lab was a lab no longer, but rather a dark room filled with broken glass and a cloud of negativity. Blaine and Tracey were there, looking sad. Anger flashed in President Lincoln's eyes. Socrates and Aristotle remained stoic. But, most shockingly of all, Cecil sat in the middle of the dark, broken mess with an expression on his face that seemed to communicate he was not at all happy to see her.

The only person in the room who'd had a smile on his face had been Paul Sims. Summer hadn't been expecting to see Paul. She didn't know exactly why her old junior high classmate was there. Regardless, he had been the one who'd heroically caught her in his arms as she zipped off the bottom of the slide, before she had crashed into any of the scattered and sharp shards.

Before she could utter a word, Paul had whisked her away, up the stairs, through a door, and back above ground. Her old friend made sure she was okay and walked with her from Cecil's house next door to Mrs. Pascapali's house. The kindly widow, who Summer had resided with for a short time before, had once again invited the scientist to stay in the guest room. Summer accepted her offer, hoping to take the time to make sense of the unexpected and uncharacteristic look she'd seen on Cecil's face.

Right now, Summer found herself sitting on Mrs. Pascapali's

back porch, drinking flavored water with Paul Sims. She wanted to be next door. She wanted to be drinking lemonade with Cecil. She wanted to know what was wrong with him and why he had looked at her like he had. But instead, here she was with Paul, who for some reason seemed extra happy. Wasn't he concerned about Cecil, too? Why was he so chipper?

Splash!

Uncle Cecil plunged down into the pool, causing a magnificent explosion of water, drenching Blaine, Tracey, and President Lincoln where they stood. Cecil resurfaced rather quickly, probably because of the pink bunny floaty around his waist. He immediately exclaimed, "Okay ofenuff yerfleanlogaffs tuseewer yulvee ziffinto firfer fizzegs!"

Dripping wet, the twins began to ask their uncle to take the snorkel out of his mouth and repeat himself. But this time they were actually both able to decipher his enthusiastic snorkel-spoken statement.

"Okay, open up your LINLOC apps!" Blaine said, looking at his sister.

Tracey then finished the waterlogged statement, "To see where you'll be zipping to first for physics!"

Without hesitation, the twelve-year-olds pulled their smartphones out of their backpacks, found them to be dry because of their waterproof cases, and then opened up the information-citing application.

"We're zipping to Papua New

Guinea!" Blaine shouted. "Longitude 9° 24' 43.39" S, latitude 147° 29' 43.26" E. Where we'll be studying the topics of forces, motion, speed, acceleration, and centripetal force!"

Tracey then added, "All with the help of a local expert by the name of Susa Dasini!"

Still wearing his over-the-top swim gear, Cecil clambered up from the pool to the pavement to join his niece and nephew. They were soaking wet, but they were already putting on their harnesses and helmets in order to ride the invisible zip lines. In sheer joy, he observed their enthusiasm as they calibrated their specially designed three-ringed carabiners to the longitude and latitude coordinates. He and President Lincoln smiled widely as the carabiners snapped shut and immediately connected to the invisible line that would whisk them away in approximately seven seconds. He waved gleefully at them as they hung a few feet in the air, waiting for the zip to happen. And then, whoosh! They were gone! Off to start their study of physics.

"Umso frouduv dosetu!" The snorkel-wearing scientist gushed to his lab assistant.

President Lincoln just sighed.

Splash!

Blaine and Tracey had landed in water in Papua New Guinea. The invisible zip lines landed them as close to their local expert as possible without being detected. The twins assumed no one had witnessed their splashing arrival, but as usual, they were momentarily stripped of sight and strength.

While traveling along the lines, the rider would soar through swirls of exciting, multi-colored light. But upon landing, the rider would be rendered weak, blind, and tingling for a few seconds until slowly but surely, strength and sight returned. The Sassafras twins were currently in this state of recovery.

Each twin felt their strength fully return, vision going from blind to white light and then finally back to full color. Blaine and Tracey had indeed landed on the water. They were floating, but not getting any wetter. The twelve-year-olds had landed in a hollowed-out log. One that was navigating its way through the exciting twists and turns of an amusement park ride. Both were about to shout out exclamations of joy when they heard approaching voices.

"A force is a push or a pull," a female voice said.

"Examples of forces are gravity, magnetism, friction, kicking, and dragging," the voice continued. "A push force can start objects moving or change their speed as the object moves away from the force. A pull force can also start objects moving or change their speed as the object moves toward the source of the force."

Looking out from their log over the water, the twins saw a wide, paved walkway. A short woman with short black hair and beautiful dark skin stood beside two other people. One was a suit-wearing middle-aged man of average height and build with graying hair and a scowl, and the other was a young twenty-something, tall woman with long blonde hair, a smile on her face, and a bounce in her step. The two women were wearing aqua-colored polo shirts with a small kangaroo teddy bear stitched on them. The man was wearing an expensive suit, complete with a button-down vest.

"Forces act on objects, causing them to move, speed up, and

slow down," the short woman continued. "Forces can also make objects change direction and shape."

"Enough with all the science talk," the middle-aged man complained in staccato-like accented English. "I simply want to know the names of the rides, the general idea of how they work, and if you think they're safe or not."

The woman, who the twins were fairly certain was their local expert, Susa Dasini, ignored the man. "Some forces act visibly by touching the object. We call these contact or direct forces. Other forces act visibly and can seem to act on the objects from a distance. We call these distant or indirect forces."

The man growled after his perturbed statement went unacknowledged.

"Forces are measured in Newtons, which were named after Isaac Newton, a famous physicist. One Newton of force causes a one kilogram object to accelerate by one meter per second." The short woman finally paused, took a breath, and then looked toward the man.

"Finally, a break from the science nonsense. Now, can we get to the ride in front of us? This, what is it called? The Kooky River?"

Their local expert pursed her lips and drew in a deep breath before she said, "As an investor in Centripetal Village, I would think that it is important to know the science behind where you are putting your money. I apologize if that is not the case. We have named the ride, Kroko River."

The Sassafras twins smiled as they watched the woman stand up for science. They were learning a lot as they listened to the conversation on the wide walkway.

They watched as the blonde woman bounced on her toes and interjected, "Did you know that krokodail means crocodile in Papuan."

"I'm not here for a language lesson, Ms. Glot," the man

scowled at the tall woman before turning to the other lady, "Or a science lesson, Ms. Dasini. We're here to assess the viability of this amusement park. And so far, I am not impressed. What even is this water ride? Does it work?"

Blaine and Tracey ducked back into their log as the man gestured wildly in their direction.

"Mr. Becker, I can assure you it works. It's our take on the traditional water flume log ride. Here at Centripetal Village, our log ride includes a double waterfall drop along with several live crocodiles," Dasini countered, clasping her hands in front of her.

Wait! The Sassafras twins glanced at each other. Had she just said LIVE CROCODILES!

The twins looked at the moving water they were floating through in a different light. What had caused that brief splash to their right? Was that a dark shadow lurking just beneath the surface to their left? The Sassafrases slunk down a little lower in their log at the thought of being surrounded by krokodail.

All at once, a loud noise sounded out from underneath Blaine and Tracey. A chomp? A crunch? No, it was a clank—the sound of a powerful motor. Their log had hit a motor-powered conveyor belt. And they were going up. Way up.

As the hollowed-out log vibrated up the steep incline, Blaine and Tracey felt a little relieved they were being lifted out of the supposed crocodile-infested water. However, they simultaneously felt worried about heading up to such a lofty point.

Over on the wide walkway, Susa Dasini answered the investor, "On the Kroko River, riders will be loaded into hollowed-out logs which float gently through serene waters at first. The logs are then taken up a conveyor belt to the height of fifty feet, where they will drop through a waterfall and land in turbulent, crocodile-filled waters on the other side. They will traverse through the jungle waters to the final waterfall, where they will drop thirty feet to the disembarkation zone. At each stage, we have the latest safety measures in place." She gestured to the blonde-haired woman next to her, "Polly can provide you with that documentation if necessary, including the results from the many simulant tests we are running."

Mr. Becker grunted. "Yes, I will need to see all of that, Ms. Glot." The three on the sidewalk looked from each other to out onto the Kroko River.

Polly's eyes found the log with the Sassafras twins and she said, "Um, Susa? I thought you said we were only running simulants on the ride so far?" Polly pointed at the Sassafras twins who gulped and waved sheepishly at the group.

"Oh no," They barely heard their local expert say over the continued clank-clank-clank of the motorized conveyor belt. Mr. Becker took a step backward with a look of horror on his face. Polly smiled joyfully and returned the twins' waves. Susa Dasini did neither—the short woman's expression looked neither alarmed nor happy.

Before anyone made a move, the twins' log reached the highest height of the conveyor belt, cresting the apex. The Sassafrases looked forward as the nose of their log changed from pointing up to pointing down, fear mirrored on their faces.

Streaming Motion

Splash!

The Chinese astronaut gratefully washed his face as he leaned over his sink.

Running water sure was easy to take for granted. That is until one had spent months in space living without this convenience.

Yang Bo wiped his face with a fluffy cotton towel and stood upright again, smiling as he looked at his reflection in the bathroom mirror. His tenure on the International Space Station had been extraordinary, but he was glad to be back home on earth, with gravity firmly connecting his feet to the ground. He was also excited about the trip he was about to take.

In a few days, he would fly from his home in China over to the United States. He planned to visit his old junior high classmates, Cecil Sassafras and Summer Beach, in the town where he had been a foreign exchange student. He was also looking forward to seeing Cecil's niece and nephew, Tracey and Blaine, again. He wanted to hear how the rest of their summer science-learning had gone. He was certain they were just about finished zipping through all the learning locations their uncle had planned for them. When the twelve-year-olds had visited him in space on the I.S.S. they had proven to be both heroic and wise for their age. He was confident the two could overcome any obstacle they faced and would succeed in anything they set their hearts and minds to.

Splash!

The Sassafras twins' log forcefully finished its first plunge and now turbulently transitioned from falling flume to raging rapids. They were freshly drenched and their stomachs were in their throats, but they'd survived . . . so far. Would they survive the crocodile-infested rapids and the final waterfall?

The three from the walkway appeared on a smaller walkway closer to where the twin's log was navigating the start of the rapids.

"Look! They made it!" Polly cried out happily.

"Harrowing! Absolutely harrowing!" Mr. Becker shouted with a look of horror still plastered on his face. The man pulled out a smartphone and began presumably filming the two in the log as they swished along.

Blaine and Tracey hung on for dear life amidst the floating

turbulence. They could actually see reptiles lurking under the surface.

The twins thought maybe "amusement" might not be the correct adjective for this ride. Maybe "harrowing" was better, like Mr. Becker had said. But really, even though this log ride was scary, it was also thrilling. Wasn't that the point of an amusement park ride after all?

In reality, the twins had felt secure in their hollowed-out log from the initial meandering, through the big waterfall drop, and even now in these rapids. It was designed well, allowing for thrills, but not forcing a fear of flipping into the crocodile-infested waters.

All at once, the rapids ended. The twins found themselves falling again. This time, they raised their hands and hollered in delight as they descended the final waterfall. All too soon, their log softly thudded into a designated disembarkation point.

Blaine and Tracey carefully climbed out of their craft onto dry land. As they did, they encountered five individuals. The three who'd been on the sidewalk were here now. Additionally, there were two slumped and disheveled mannequins, similar to crash-test dummies, leaning against a nearby railing. The poor things looked more than a little battered and beaten up. One had a cracked torso and the other was missing an arm.

"Well, hello you two!" the tall, blonde-headed woman said, skipping up to the twins with a smile. "My name is Polly. Polly Glot. I'm going to serve as a translator here at the park, once it opens. And you guys are our amusement consultants, right?"

"Well . . .we—" Tracey started to say, but was cut off by the happy, energetic young woman.

"Good thing you two are here today." She laughed and pointed at the crash-test dummies. "Because our previous amusement consultants really need a break!"

"Umm . . . if they're like that . . ." Tracey fumbled, still trying to put together a sentence. "Then should we really be . . ." Tracey wasn't able to finish before Polly interrupted again.

"This here is one of my heroes!" Polly said, referring to the local Papua New Guinean. "Susa Dasini is our wonderful leader here at Centripetal Village." She then gestured with less enthusiasm toward the gentleman. "And this is Bavo Becker. He is a potential investor."

Susa acknowledged the twins with a wave. Bavo did not.

"After a whole lot of dreaming, planning, building, and perfecting we are almost ready to open this amusement park," Susa Dasini said. "We just need a couple more investors to help us with some final finishing touches. I was under the impression that Mr. Becker was likely to be one of those investors, but that might not be the case. We have spent the majority of today's tour proving to Mr. Becker that the park is safe, instead of talking about finishing or opening it."

"Safe?" Bavo said, a vein bulging in his forehead. "From my perspective there is nothing safe at all about this park! Live crocodiles on a log ride? Kids serving as 'amusement consultants'? Preposterous!"

Polly ignored Mr. Becker and smiled at the twins. "Thank you, you two! Thank you for being the first actual people to ride the Kroko River!"

"That's right, Polly," Susa said. "We knew these two would be safe thanks to our extensive research into forces and motion. If you remember, ancient scientists were puzzled by motion. Aristotle thought that an object moved when a force acted on it, but when the force was removed, it stopped. However, this didn't explain objects that flew through the air on their own."

"Here she goes again," Bavo said under his breath with a roll of the eyes.

Ignoring his underhanded comment, Susa continued sharing

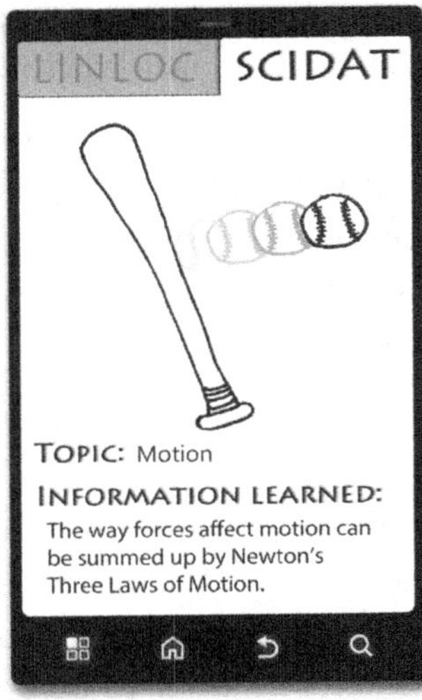

scientific facts. "Many years after Aristotle, Galileo said that a force was needed to start, stop, or accelerate an object, which was much better but still didn't give the complete picture. Then Isaac Newton came along and built upon Galileo's ideas to come up with the three laws of motion we still use today. Isaac Newton studied at Cambridge, but it was while he was home because of the plague that he discovered the Three Laws of Motion and a type of math we know as calculus. He also explained how gravity keeps the planets orbiting around the sun. But regarding the topic at hand, how forces create motion, it can be summed up by Newton's Three Laws of Motion as follows . . ."

Mr. Becker blew out a frustrated sigh as Susa grabbed a ball from a nearby cubby on the platform. The twins listened, riveted, as their local expert demonstrated the laws of motion. She set the ball on the railing and said, "An object in motion will stay in motion and an object at rest will stay at rest unless an outside force acts on them." Susa pushed the ball gently, and it moved slightly before it stopped.

"Second, the greater the force of an object, the greater the change in its motion. The greater the mass of an object, the greater the force needed to change its motion." Then, she pushed the ball harder, and it rolled down to the post, hitting it and bouncing back along the railing toward them as she said, "And finally, for every action, there is an equal but opposite reaction."

"Enough! I've had enough!" Mr. Becker interjected. "I didn't travel all this way for a science lesson! I simply want you to tell me the names, general idea, and safety levels of these rides. That is all I care about."

"Mr. Becker," Polly attempted to say with a smile of defense. "I think Susa is just trying to explain the physics behind the rides to help you understand the safety fea—"

"I said I've had enough!" The man cut Polly off. "I'm going to go call Alwin. Another potential investor, who also happens to be a safety inspector with great authority in this region. I'm going to fill him in about what's happening around here! I'm sure he will have some strong opinions about all of this nonsense."

"There's no need to call Mr. Aldrich," Polly said as the man in the suit turned and stormed off.

The Sassafras twins weren't sure how to react to the frustrated man's exit. The two women chose optimism.

"Well." Polly turned back to the twins and their local expert. "I would like to hear about the rest of the science behind the park. How about you two?"

Blaine and Tracey nodded, and Susa continued. "When talking about forces and motion, one should be familiar with inertia and momentum. Inertia is the tendency of an object to resist motion. The more the object weighs, the greater the force it takes to get it moving. Momentum is the tendency of an object to keep moving. The more an object weighs and the faster it goes, the more momentum it has, meaning it will be harder to stop."

As the Papua New Guinean finished, the blonde-headed Polly clapped her hands, "Thank you, Susa Dasini!"

The Sassafras twins awkwardly clapped with her until Susa turned to them. "What an awesome maiden voyage on the Kroko River! Great job, you two! And thank you again for serving as our new amusement consultants!" Polly said with a smile.

Susa added, "As you can see, it's probably past time we retired those two." The young woman pointed at the crash-test-dummies.

Blaine and Tracey shivered, feeling glad they hadn't finished the ride looking like that.

"I just love this place!" Polly explained. "I had the wonderful privilege of spending a couple of years of my childhood here. With my dad being in the military, our family moved around to a bunch of different countries. Totally fine by me because I love all the different cultures. There's something about being able to ask for coffee in multiple languages that is so exciting! My favorite of all the countries we ever lived in was Papua New Guinea, so when I finished college and launched out on my own, I moved back here to reunite with my old friends Susa, and now I get to be the lead translator for Centripetal Village! Not that you two asked for my life story!"

Blaine and Tracey shrugged their shoulders.

"Stop over sharing, Polly," she whispered to herself before glancing back up at the twins. "Enough about me. I haven't even asked what your names are yet.

"We are the Sassafrases," Blaine answered for himself and his sister as he often did. He liked to assume the role of older sibling, even though he had only been born five minutes and fourteen seconds earlier than his twin. "Blaine and Tracey Sassafras. We are excited to be in Papua New Guinea and here at Centripetal Village, but we are just as excited for the science of the rides."

"Well, that's perfect," Polly Glot exclaimed. "Because my boss here, as it so happens, is the most brilliant person I know. And she loves science, especially physics!"

"It sure sounds like it," Tracey said. "We would also love to hear anything and everything you want to share about this amusement park."

"Several years ago," Susa said. "A group of us set out to establish the first ever amusement park in Papua New Guinea. We had the know-how and the science knowledge to pull things together. With our connections, we had the indigenous manpower to commit to the project. I'm proud to say that the leadership team—the foremen, architects, and engineers—are all native to

Papua New Guinea. But one thing we did need outside help with was investment capital."

Polly jumped in, "That's where I come in! I head up the Centripetal Village investment campaign, working on building a partnership with a group of investors. To our delight, investors from all over the world hopped on board. We couldn't have gotten to where we are today without them. So far, we've raised ninety percent of what we need, and we're hopeful that we can raise the last bit with a few more investors. Everything was going well. But then . . ." she paused and looked in the direction Mr. Becker had walked.

". . .that guy showed up. At first, he seemed like a legitimately interested investor. However . . ." The blonde paused again. "Well, let's just say he doesn't have a firm belief in the local workforce, and he isn't sure Susa should lead the project."

The twins' faces scrunched up a bit in confusion. They didn't understand why he would feel that way. Especially with Susa, who seemed so smart and capable.

"He started sowing seeds of doubt among the new group of potential investors," Polly explained. "He proposed the idea of having outside engineers and architects come in and take the lead in the completing of Centripetal Village. He thinks the local men and women don't really have what it takes to finish such a large project or to operate things safely." Polly frowned and shook her head.

Susa stepped back in. "That's why I insisted on sharing the science behind the rides. He needs to know we didn't just slap these things up. That we understand the physics at work in our rides. And knowing that, we have done everything to make them as safe as possible."

"And now, because of his nay-saying, we may not be able to raise the final ten percent we need to open Centripetal Village," Glot said with down-turned lips, but she didn't stay that way.

"If he's so down on the park, why is he here today?" Tracey mused.

Polly and Susa shrugged their shoulders as a silence descended over the four.

A silence that was broken by the sound of angry footsteps. Before they had time to react, Mr. Becker shouted, "I'm here to shut the park down!"

CHAPTER 3: CONSULTING IN THE CENTRIPETAL VILLAGE

Amusement Acceleration

"Before you shut us down, Mr. Becker, how about we show you what else the Centripetal Village has to offer?" Polly plastered on her sweetest smile.

Mr. Becker folded his arm across his chest, scowling. The other hand still held his phone to his ear.

"Give us a chance to convince you how wonderful this park is," Polly said.

Mr. Becker tipped his head up in an I-am-listening gesture.

"Shall we go to the Ring of Paia next?" Polly moved her arms to show which direction to go. The group filed in behind her, but Becker kept his distance.

"The Ring of what?" Blaine asked curiously.

"Paia," Polly Glot said again.

"Paia?" Tracey asked in the same fashion as her brother.

"Yep, *paia*," Polly confirmed. "It means 'fire' in the local language."

"So the name of this roller coaster is the Ring of Fire?" Tracey gulped.

"That's right," Glot smiled. "The coaster is named the 'Ring of Paia' to commemorate the fact that Papua New Guinea sits in the heart of the Pacific Ring of Fire."

Blaine and Tracey remembered learning about the Ring of Fire from when they studied geology. It was named for the volcanoes and earthquakes that frequently occurred around the edges of the Pacific Ocean. They weren't sure they wanted to revisit

this ring in roller coaster form.

The group entered a structure that was built to look like a volcano, walking over to a platform where a line of sleek coaster carts awaited.

"This awesome coaster will be known for its speed and acceleration. It shoots riders straight out of this recreated volcano, picking up tremendous speed, and then sails through several rings of fire!" Polly exclaimed excitedly.

The Sassafras twins shuddered with tinges of fear, but hid it all with plastered-over smiles. The reluctant yet excited new amusement consultants took their places side by side in the front car. As the siblings strapped themselves in, they pushed the images of the beaten and torn crash-test dummies out of their minds. They sat still, staring ahead at the track.

"Rings of fire, huh?" Blaine chuckled nervously, preparing to be shot out.

"Speed and acceleration?" Tracey followed suit, eyes scanning the opening above them.

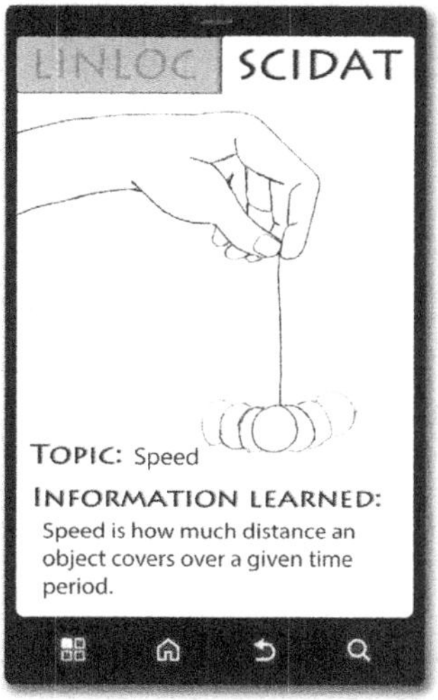

Susa Dasini nodded as she heard the words speed and acceleration. "Speed is how much distance an object covers over a given time period," their local expert said. "When measuring speed, average speed is used because some objects, like vehicles for instance, can speed up and slow down in small amounts over the measurable time period. Instantaneous speed, like the speed measured by a speedometer, is the speed an object is traveling at a given moment. Speed in a particular direction is called velocity."

"Velocity, huh?" The twins asked in unison, grips tightening on the metal handles of the safety bar.

Polly and Susa nodded gleefully. On the other hand, Mr. Becker, the European investor, continued scowling from a distance. He stood next to the rest, yapping pessimistically on his phone. Would the two of them really be able to shut down Centripetal Village before it even opened? Blaine and Tracey didn't know.

"Acceleration is when an object changes speed," Dasini continued with the scientific information. "Technically, it is called acceleration when the velocity is increasing. It is called deceleration when the velocity is decreasing. Acceleration happens when forces are unbalanced."

"I'll tell you what's unbalanced!" Becker said, rejoining the group but not ending his video call. "The safety protocol at this atrocious park! I mean, look Alwin, just look! Susa has these two kids strapped into this roller coaster. Sure, their safety belts look secure enough, but I just overheard them say the coaster is going to shoot through rings of fire! Literal fire, Alwin! First it was the crocodiles and now they're flirting with flames!"

The investor swung his smartphone around, presumably so Aldrich could see what he had shared. As the phone bobbed left and right and up and down, the Sassafras twins could see the inspector's face on the screen. He appeared to be scowling, just like Mr. Becker.

Paying no attention to the investors, Susa and Polly walked over to the control panel for the coaster.

"Okay, Blaine and Tracey," Polly said excitedly. "Are the two of you ready to test out the Ring of Paia?"

The twelve-year-olds took quick but deep breaths. Blaine looked over at his sister. "Velocity," he stated.

"Speed and acceleration," Tracey returned with eye contact.

"Literal rings of fire," they said together, still looking at each other. They both then turned toward the two women. "We're ready!"

Immediately and enthusiastically, Susa pushed a large button on the control panel. The twins expected to be immediately shot out like cannonballs from a cannon. But the coaster slowly lurched forward, taking them from the covered loading area out into the open. Once all the cars of the linked train were under open sky, the coaster came to a stop. Looking out ahead from their seats in the first car, all the twins could see was a straight flat and fairly boring-looking length of track jutting out a short way in front of them. They didn't see any rings. They didn't see any fire.

Next, some kind of loud and large weights and balances engaged somewhere below and began turning the Sassafrases and their coaster. The short section of the track they were occupying flipped around. Once the 180-degree turn was completed, the twelve-year-old duo could see that 'fairly boring' was no longer applicable to the track ahead. And there were rings. Lots of rings. With fire.

"Okay, Blaine and Tracey, hold on tight!" Polly shouted, still standing next to their local expert near the control panel. "Susa is about to push the next button, which will engage powerful and compressed air that is going to launch you forward from zero to sixty miles per hour in less than three seconds. You'll zip back through the building, and then up and out through all the rings of fire! And, oh, I think there are several upside-down loops too!"

The twins sucked in deep breaths of anticipation. Suddenly, *click, zip, bang,* and off the Sassafrases shot. Their heads pushed back firmly against the headrests of the car. Blaine and Tracey felt like fired cannonballs.

Back through the loading area they rocketed, glimpsing two waving women and one scoffing man. They reached open track and top speed. Their stomachs raised into their throats. Their cheeks stretched to wind-forced pliability. The roller coaster flew right through the center of the large flaming rings of fire. The coaster banked and sped through several upside-down rings.

The brother and sister involuntarily screamed as they went from right-side-up to upside-down and back again several times in a matter of seconds. The heat from the surrounding fire reached them without scorching them.

After several more minutes of loops, banked turns, and rings of fire, the coaster slowed down and found its way back to its starting point, coming to a complete stop. Blaine and Tracey let out whoops of excitement. Polly and Susa clapped, while Mr. Becker continued scowling into his phone.

"I mean, there's no question now. We've got to shut this place down. Am I right, Alwin? First the crocodiles and now the fire! This place is just too dangerous to even think about letting it open. We can't submit would-be visitors to this kind of insanity!"

On the other end of the call, Alwin Aldrich could be heard grumbling softly. The Sassafras twins stepped out of the coaster car and joined the three adults as Mr. Becker said, "Alwin, we have to shut this place down. Centripetal Village should never be allowed to open its doors! We must sto—"

Suddenly, out of nowhere, some kind of a furry ball of energy bounded up, knocking the phone right out of the investor's hands. The phone flew a couple of feet and then hit the deck of the roller coaster's loading area. The device slid to a stop without a crack or scratch on the screen. The twins saw Mr. Aldrich with a raised eyebrow.

"This is all just too much! Too much, I say! This park is already dangerous and absurd, and now this! What is that thing?" Mr. Becker asked as he pointed at the animal.

"Matschie"" Susa Dasini answered.

"Matschie?" The investor asked as he moved to pick up his phone.

"Matschie," Susa said again in confirmation.

"What is it?" Tracey asked in delight as she and her brother gleefully regarded the cute, fuzzy animal that looked like a mix between a small kangaroo and a stuffed teddy bear, or maybe more like the adorable red pandas that Tracey had seen at the beginning of the summer at the Panda reserve in Sichuan, China. The animal looked similar to the one stitched on the women's shirts.

"Matschie," Polly confirmed the name of the animal. "He's the official mascot of Centripetal Village!"

"Official mascot? Sick!" Blaine approved with a nod.

"Matschie is a tree kangaroo," Polly continued. "Tree kangaroos are a rare species of marsupials found here in Papua New Guinea and only a few other places in the region."

Polly bent down to pet the three-foot-tall marsupial's auburn and white head. "Susa thought this little guy would be a perfect mascot for our amusement park."

The Sassafras twins agreed as Tracey said, "He's so cute!"

"Cute?" Mr. Becker shouted in defiant disagreement. "This

thing looks like a giant rat! Vermin like this shouldn't be uncaged, free, and openly roaming about. It's ludicrous!"

"Well, Matschie is usually very friendly. He is only ever aggressive when he feels threatened. He was raised in—" Polly started but wasn't able to finish before being interrupted.

"Alwin! Alwin! Do you see this?" Becker picked up his phone and pointed the camera toward the tree kangaroo so that his counterpart could see. "We have to shut this park down, Alwin, and we have to shut it down now!"

Blaine and Tracey were growing more than a little tired of Mr. Becker and his negativity. They had initially felt trepidation about this park, especially after seeing the state of the crash-test dummies. But after safely experiencing two rides, they were feeling much better. Meeting Polly and Matschie, along with learning science from Susa, helped confirm that the park was safe and fun. Sure, there were mind-bending and gut-busting thrills, but that's what an amusement park should have, right?

The twins suspected their local expert, Susa Dasini, was even smarter than she let on. They guessed the park and all of its rides, down to the smallest detail, had been planned perfectly by her and her team members. But the potential investor in front of them seemed to miss all that. He continued to lambast the park to his friend on the phone. He couldn't see how absolutely amazing the Centripetal Village would be.

As Mr. Becker continued to rant and rave, Susa picked up Matschie, hugging the animal to her. She approached the Sassafras Twins. With a smile and a point, she asked them, "Do you two want to go ride the Merry-go-Matschie? It's a more low-key ride that is designed for young thrill-seekers. But it's still loads of fun."

"Absolutely!" The Sassafras twins answered emphatically.

It wasn't skipping. It wasn't dancing. And it definitely wasn't any kind of be-bopping. It might, however, be sauntering. Yep, he was pretty sure that's what it was. He was fairly certain he was sauntering.

Usually he be-bopped his way out to the mailbox to check to see if he'd received any envelopes, flyers, or small parcels, but today he sauntered. It wasn't exuberant sauntering, and it wasn't doleful sauntering. It was somewhere in between.

Cecil side-stepped a small puddle and pulled open the mailbox door to see if anything was inside. "Oh willy nilly, Tom and Billy, we have ourselves an envelope," the red-headed scientist said out loud even though he was all by his lonesome.

The envelope was upside down. Cecil reached in, grabbed it, and flipped it over and read what he found: "Cecil Sassafras. 1104 North Pecan Street . . . but hmmm, humminy hum . . . no return address."

The scientist tore into the envelope right there on the street, finding an invitation to a reunion inside. It said:

> Dear Cecil Sassafras,
>
> You are cordially invited to a junior high class reunion!
>
> Come back to the junior high school building and enjoy getting reacquainted with old friends and faces.

The high-quality invitation then listed the date, time, and even the classroom number where the reunion would be held.

Cecil smiled. Maybe this was just what he needed to eliminate the slow sauntering and put the be-bop back in his step: a junior high class reunion! It would be great to see all his old pals and classmates again. The scientist's brain raced with excitement, but then it, like his stride to the mailbox, suddenly slowed to a saunter.

Summer Beach was an old junior high classmate of his. Would she attend this reunion?

Centripetal Center

The Merry-go-Matschie had been a blast. Yes, it had been a little more low-key than the Kroko River and the Ring of Paia, but it still had been loads of fun. Blaine and Tracey had ridden side by side in a gondola designed to look like a tree kangaroo. Strong mechanical arms connected the gondolas to a large spinning center. The center looked like a big, thick tree trunk with arms cleverly designed to look like tree branches. The Sassafrases went around and around and up and down, feeling like small children again.

Though the Merry-go-Matschie ride had no apparent safety issues at all, Mr. Becker had still found plenty to complain about. The group had then moved to a giant swing, a Ferris wheel, and even bumper cars. All the well-designed rides had included connections to Papua New Guinean culture and employed the laws of physics.

"Okay, Blaine and Tracey," Polly exclaimed with the most excitement in her voice yet. "We have now arrived at the very center of Centripetal Village! The ride that towers before you now is the crowning achievement of the entire amusement park!"

The Sassafras twins were enjoying their job as amusement consultants more and more. They had caught glimpses of the cone-shaped tower as they had walked through the rest of the park. But when Polly pointed it out, they craned their heads back to take it all in.

"Behold, the Colossal Cone!" Polly said as she gestured boldly forward with her arms, presenting the gargantuan ride.

The twins appreciated the magnificently giant coaster. Their hearts beat faster at the thought of riding it.

"This coaster starts at ground level," Susa explained, gesturing with her hands. "Then you clickity-clack up at a forty-five degree angle on the track wrapped around the outside of the cone. The coaster car reaches a height of one hundred and fifty feet above the park. The track sits at the brim of the cone for a few seconds, tilting inward so you get a glimpse at the depth of the cone before plummeting down. That's when you see the hidden opening at the base of the cone, which allows the coaster to continue an additional fifty feet underground!"

The translucent surface of Colossal Cone revealed the track of the roller coaster inside and out so the twins could see what Polly was talking about. Way up high at the top, the track was a wide circle that turned into smaller and tighter circles as it dropped into the ground, giving the humongous structure its striated appearance.

Polly continued her explanation, "After the pause and the tilt, the magnetic brakes release, and the coaster winds slowly around the very top circle. As you drop further into the cone, the circles of the roller coaster get smaller and tighter. Centripetal force grabs ahold of you, and the speed will blow your mind!"

The twins felt like their minds were already blown, and they hadn't even gotten on this thing yet.

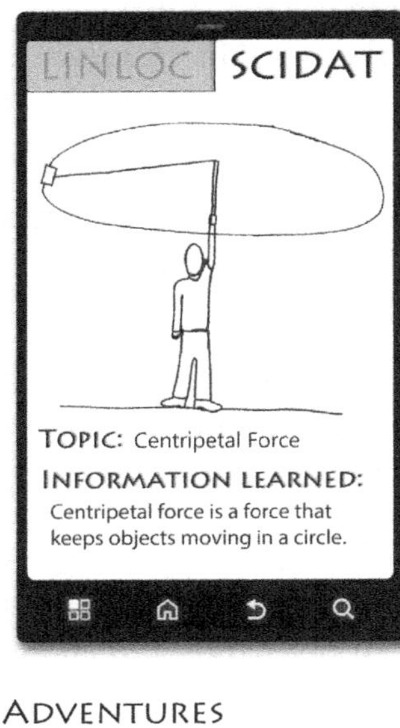

"Centripetal force is a force that keeps objects moving in a circle," Susa Dasini suddenly cut back in, prompted by the mention of Blaine and Tracey's next scientific topic. "Merry-go-rounds, wheels, and spinning tops all go around in circles. All of these experience a force that pulls the objects in toward the center of the circle. We call this centripetal force."

"Gyroscopes are devices that purposefully use centripetal force," the woman explained. "They have spinning wheels that resist the pull of gravity. Gyroscopes can be simple or electrically driven. Electric ones are sometimes used to help boats stay upright in rolling seas."

Susa looked from the twins up at the giant cone the twelve-year-olds were staring at. "And of course, roller coasters experience centripetal force. It is a centripetal force that holds a person onto a coaster and the coaster cart on the track as it goes around in loops. It feels like something is pushing the objects up, but in reality, this is centripetal force. In this case, it is using inertia to try to make the coaster cart and its passengers go straight ahead."

"Centripetal farce is more like it!" Mr. Becker blurted out. "This giant eyesore in front of us is a joke. And I will not let you put these poor children on this ridiculous cone coaster. I'm also not going to help you put anyone on any of these rides. Ever. I'm pulling the plug. Right now. I'm shutting down Centripetal Village once and for all. The rides are poorly designed and terribly unsafe. I'm officially petitioning to shut it down right now!"

Was he serious? Was he really going to shut this park down? Did he even have the authority to do that? Was he really going to pull the plug on Centripetal Village?

"Isn't that right, Alwin?" the investor said to the man on his phone. "It's time to shut this thing down. You're with me, right?"

"I am with you," the inspector's voice could finally be heard clearly. "And when I say with you, I mean I am literally with you."

Wait, Alwin Aldrich's voice wasn't only coming from the smartphone. It was also coming from behind them!

The group did a 180-degree swivel on their heels, seeing Mr. Aldrich live and in person. He was tall, like Mr. Becker. He was blonde, like Becker. He was wearing a suit, like Becker. And he was on a phone call, like Becker. A phone call he hung up on once everyone faced him.

Mr. Becker seemed just as shocked to see Mr. Aldrich as Polly,

Blaine, Tracey, and Susa did. Matschie, however, was unreadable.

"Yes, I am with you in person." Mr. Aldrich said. Then he paused, looking everyone in the group in the eye. He landed on Mr. Becker and added, "But I am not with you in your decision."

"What do you mean?" Bavo asked, confused. "Of course you're with me on shutting down this park. Haven't you seen all the safety violations?"

Blaine and Tracey didn't know what to do but stare at the two men. Even Polly and Susa were quiet and still. Matschie, however, made an animal sound and suddenly jumped directly at Aldrich. The tree kangaroo landed directly in the investor-inspector's arms. Mr. Aldrich patted and rubbed the marsupial's head and behind his ears like a beloved pet.

"What are you doing, Alwin?" Mr. Becker asked, his voice rising several octaves. "Why are you petting that vermin? And what are you even doing here? I thought you were in Germany. The whole time we were on the phone, I thought you were in Germany."

"You thought wrong, old friend," Aldrich answered evenly. "As I said, I am with you in person, Bavo, but I'm not with you in your decision. Not only do I not want this wonderful amusement park to shut down, but I want to put my official stamp of approval on it right now. Additionally, I would like to contribute the final ten percent to get the wonderful place completed. Let's get Centripetal Village open to the public!"

"What?" Becker barked, befuddled.

"That's right," Aldrich confirmed. "Open the gates."

"But, but, what about that crocodile log ride?" Mr. Becker bumbled. "You can't approve that. It's not safe. You can't allow people to float around in water with live crocodiles."

"The crocodiles are live," Mr. Aldrick nodded. "But they are not swimming in open water next to the logs. They are cleverly housed behind a disguised and reinforced plexiglass wall. It's a great habitat for the reptiles, and a thrilling experience for the riders."

"Okay, well, well, what about that coaster covered in fire?" Becker bemoaned. "That thing is deadly."

"It absolutely is not," Alwin resolutely disagreed. "The Ring of Paia coaster actually contains no fire or open flames."

"But I saw it!" Mr. Becker bellowed. "You saw it! We all saw it!"

The investor-inspector shook his head. "Every exciting looking flame you saw was an illusion. They were created with light shining against moving pieces of fire-colored plastic plus a blast of hot air."

Mr. Becker refused to accept his colleague's words. "What about the Merry-go-Matschie and the—"

"All safe," Mr. Aldrich interrupted before the man could go on. "It's all safe. It's all fun. And you know it."

"But what about her?" Bavo blubbered, pointing at Susa Dasini. "She is so frustrating, and . . . and . . . she obviously despises me. She won't stop talking about physics! Physics, Alwin?"

At this point, Alwin Aldrich seemed to be amused by his counterpart's silly exasperation. The twins thought maybe the investor-inspector might even be chuckling. "Susa Dasini is brilliant. And she definitely knows her physics. Just look around. Centripetal Village, with all its rides and attractions, uses physics and follows its laws to create a wonderful world of amusement. She's done it—she and her teammates. They have built Centripetal Village without us, and it's time for its gates to be open! It's time for us to partner with them to get the word out. I wholeheartedly give this incredible place my official stamp of approval, and you should too, old friend!"

For the first time all day, Mr. Becker didn't spout off a rebuke, rebuttal, or complaint. His scowl remained, but his arm posture relaxed.

"It's time for you to stop being angry, friend. Instead, start being thrilled. How about you and me ride this roller coaster? How about if we all ride it? What do you say?"

Several minutes later, Becker was the only one screaming, but he was not alone in feeling the thrills. Mr. Becker held both his hands high up in the air. He had a humongous smile on his face, accompanying his joy-filled scream. The investor was having the time of his life riding the Colossal Cone roller coaster. All of his companions were also having a wonderful time. Alwin Aldrich rode happily next to his friend seated on the front row. On the second row, Susa Dasini and Polly Glot sat next to each other, thoroughly enjoying the ride with laughs and smiles. Behind them, Blaine and Tracey had their arms fully extended in blissful exhilaration as the coaster went faster and faster, dropping into the tighter circular curves of the cone. On the very back row of the train sat the original amusement consultants, the crash-test dummies. Blaine and Tracey weren't sure how they had gotten on the roller coaster, but here they were.

The twelve-year-old twins released happy screams of their own as the Colossal Cone roller coaster accomplished its most extreme level of centripetal force yet. It zipped around in tight circles, dropping the riders down into the delightfully intense underground portion of the track. Blaine and Tracey were sure this was the coolest roller coaster they had ever ridden. They were also sure Mr. Becker shared this sentiment. The Sassafrases had no doubt the investor, would now be more than happy to add his stamp of approval to this park, along with his financial contribution.

Susa Dasini and her fellow Papua New Guineans had done it! They had dreamed up, designed, engineered, and built this amazing place all on their own. It would officially be Papua New Guinea's first amusement park when the gates of Centripetal Village soon swung wide open for one and all!

CHAPTER 4: CAMP ZIP FIRE

Gripping Gravity

Later in the afternoon, the science-learning siblings stole away to a quiet corner of Centripetal Village to enter all their SCIDAT data and pictures for the four topics. This had been one of their favorite face-to-face learning experiences yet.

Before they stepped away, Mr. Becker delivered his official stamp of approval. The investor had not only enthusiastically endorsed the park, he had also promised to begin financially contributing to the park and soliciting others to do the same. Susa had hugged him, and Polly started sharing their marketing ideas.

With all the SCIDAT now successfully entered and digitally sent to their uncle's basement lab, the Sassafras twins opened the LINLOC applications on their phones. When the twelve-year-olds saw the next location, they were utterly astonished. They had never been more shocked to see a line location.

Could what they were seeing really be true? Was this a joke or some kind of mistake?

Their minds wandered all the way back to the beginning of the summer—back to the backseat of that hot and shaky bus. The backseat where the two of them had sat bemoaning the fact they were heading to their uncle's to study instead of heading to their favorite place in the entire world: Camp Zip Fire.

They glanced back down at their phones. Shockingly, LINLOC was telling them they were going to Camp Zip Fire next.

They weren't sure if Camp Zip Fire was still their favorite place in the entire world anymore. They had seen some pretty spectacular locations around the globe this summer. But they were excited to visit this wonderful place. They had been going to summer camp there since third grade. A place where all their best friends went. A place where they were known and appreciated. A place where they were each a part of record-setting camp tribes. Camp Zip Fire rocked, but should they be zipping there for science? Should it really be one of their science-learning locations?

"Camp Zip Fire in Big Sky, Montana," Blaine read aloud what they were both seeing on LINLOC. "Longitude 46° 59' 37.39" N, latitude 114° 0' 57.13" W. Where we will study the topics of gravity, friction, balance, and buoyancy."

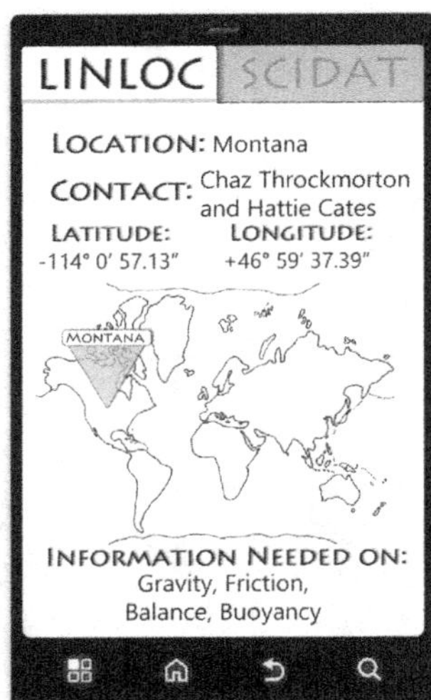

"And, oh my cow, Blaine, look at the local experts' names," Tracey said, truly surprised. "Chaz Throckmorton and Hattie Cates."

Both twins were wide-eyed and drop-jawed. Not only were two local experts listed, which was highly unusual, but the two people listed were the Sassafras' favorite camp counselors.

"Can Chaz and Hattie really be our local experts?" Tracey asked.

Blaine just shrugged. He didn't really remember scientific knowledge being an outstanding feature of either of these beloved counselors.

The brother and sister sat in their far corner of Centripetal Village for a few seconds, kind of staring at the screens of their smartphones, wondering if it could really be true that they were about to zip to Camp Zip Fire. Then, all at once, the excitement rose. They were about to zip to Camp Zip Fire!

"Tracey! We're going to Camp Zip Fire!" Blaine exclaimed.

"Yes we are, bro, yes we are," Tracey returned.

They put their helmets on and cinched up their harnesses. The specially designed three-ringed carabiners were calibrated. The correct invisible zip line located and attached to. The seven customary seconds passed. And then, zip bang, Blaine and Tracey Sassafras zipped from Papua New Guinea to Big Sky, Montana at the speed of light!

Through swirling and twirling light, they traveled and then landed in their next line location, devoid of strength and sight. As they waited for these faculties to return to normal, both twins wondered where on the familiar campground they'd landed. They pretty much knew the whole place like the back of their hands, so they were fairly certain they could recognize their landing spot.

When they could see again, that did not turn out to be the case. Not only had they arrived at night, in the dark, but they had also landed on some kind of large inflatable, vinyl surface that was proving to be very difficult to move around on. Not only could neither twin stand up, the cushiony surface seemed to be surrounded by water.

"What in the world is this, Trace?" Blaine whispered,

somewhat flustered at not being able to get to his feet. "Did we land on some kind of waterbed?"

Tracey chuckled at her brother and then took a longer moment to let her eyes not only adjust from the post-zip blindness, but also to the nighttime darkness.

"I bet it's a blob," she said.

"A what?" Blaine asked.

"It seems like we landed on a blob."

"A blob?"

"Yeah, you know, one of those big inflatable things that floats on the water and launches someone in the air from one end if somebody jumps onto the other end."

"Oh, yeah. I know what you're talking about. But wait, Camp Zip Fire doesn't have a blob.

"Well, it sure looks like they do now. Blaine, I think we landed down at the waterfront on a brand-new blob."

"Well, why didn't we launch each other up in the air and into the lake?"

"Maybe because we landed together in the center?"

Blaine nodded at Tracey's summation. With his eyes fully adjusted, he looked around and saw water all around, except behind them, where there was some sort of large wooden structure that looked like a two- or three-story dock.

Blaine carefully maneuvered his body around and then began crawling on his hands and knees toward the wooden structure. Tracey followed. In less than a minute, both Sassafrases crawled safely from the blob to the multi-level dock without ending up in the water.

The twins were glad to be on stable ground, but what were they supposed to do now? Camp Zip Fire had a rule that campers couldn't be outside at night without counselors or camp staff. "Well, I guess the best option is to go find our tribe's cabins and see if we can get in," the male Sassafras suggested.

Tracey thought her brother was probably right. The only thing was, this meant they would part ways for the time being. A creek separated the part of Camp Zip Fire where campers slept—the boys' cabins on one side and the girls' cabins on the other. Splitting up wasn't the best option for learning science in tandem, but it seemed better than sleeping outside. They might get in trouble if they got caught, especially if they got caught by Grizz.

The bearded and burly three-hundred-pound Zip Fire staffer was kind of a jack-of-all-trades. He did anything and everything required to keep the campground up and running. He was a nice enough guy, but he was also a stickler for the rules.

As quickly and quietly as they could, the Sassafras twins made their way from the waterfront to the lodging area. When they got to the creek, they gave each other a fist bump and then took the different paths to their respective sides.

"See you tomorrow," Tracey whispered to her brother.

"You, too," Blaine whispered back.

The Sassafras boy crept forward with quiet steps, passing a few cabins until he finally got to one with a sign over the door that read Mustangs. He reached out to grab the door handle, giving the knob a gentle turn. It opened. The Mustang cabin was unlocked. He stepped from the darkness outside into the even darker interior of the cabin.

He could barely make out the dozen or so bunk beds filling the room. He could faintly see the outline of the bathroom door at the other end of the space. All was quiet except for a few different tones of light snoring. Blaine quietly shut the cabin door behind him and then tiptoed over to the grouping of bunk beds to his right. Dimming the light of his phone with his hand, he found an empty bottom bunk. As stealthily as he could, he took off his helmet and his backpack, slid them under the bed, and sat down on the mattress.

A big squeak and creak rang out. He froze for a few long moments, waiting for everyone to wake up and the overhead lights to flash on. But neither of these things happened. All remained the same. With a few smaller squeaks, Blaine lay all the way down. Now safe and sound inside his cabin amongst friends. What would they think when they saw him in the morning?

Tracey made a few squeaks and creaks before getting into her cabin and into her bunk, but she was certain not a soul had noticed her arrival. She had been extremely relieved to find the Appaloosa cabin unlocked. As she closed her eyes, she wondered what would they think when they saw her in the morning?

Moments later a sound awakened her. Light streamed into the cabin. How had hours of sleep felt like only seconds? The loud cry of a bugle playing Reveille pushed her to wide awake. She sat up in her bed while all the other girls around her yawned, stretched, and lumbered up slowly. The second the bugle ceased, Tracey heard her name in question form and then in exclamation form.

"Tracey? Tracey!" It was Joanna in the bottom bunk right next to her. "Hey everybody, Tracey's here!" Joanna, Tracey's always happy friend exclaimed, now as fully awake as Tracey was.

Three heads immediately popped down from the top bunks above her. "Tracey!" Janna, Marci, and DeeDee squealed in unison.

"When did you get here?" Janna asked.

"Last night," Tracey said.

"Last night? Where have you been the rest of the summer?" Marci asked. "We've missed you like crazy!"

"Yeah!" DeeDee added. "And we need you to help us beat the Jennet tribe and set a record on the obsta—"

"Good morning, ladies!" An authoritative and cheerful voice interrupted the reunion. "You've all heard the bugle's call. That means it's time to get up and get ready. We need to be at the flagpole in thirty minutes."

The voice belonged to Hattie Cates, the beloved counselor of the Appaloosa tribe. She was smart, athletic, and pretty, but even better than that, she was courageous and kind. And somehow, someway, even though she had surely just woken up, she looked like she'd already been up and ready for hours.

The girls of the Appaloosa tribe followed Hattie's beck and call. They were ready and at the camp's flagpole with a couple of minutes to spare.

As the Mustang tribe approached the flagpole, Blaine saw that Tracey and her tribe were already there, waiting and standing at attention. He gave her a slight wave as he and his buddies, his tribe mates, and his camp counselor, Chaz Throckmorton, reverently approached. Chaz was disciplined yet cool, strict yet fun—every member of the Mustang tribe loved him. He had probably been up reading or working out long before the bugle had sounded. On the other hand, Blaine was pretty sure it had only been about five minutes ago that he and his fellow tribe members rolled out of bed. No matter—they were here, on time, albeit with groggy eyes and sleep-lines on their faces.

Upon waking up, none of his friends were acting like it was

strange he had basically appeared out of nowhere. They all just seemed to be happy he was here at Camp Zip Fire. Especially Jamal, Jarod, Matt, and Davy.

Once all the tribes arrived and were situated around the flagpole in their designated spaces, the big, burly Grizz walked forward with a perfectly folded triangular American flag. Without any of the campers from the dozen tribes making a single peep, he connected the flag to the flagpole and honorably hoisted the stars and stripes up to the top. Then, in concert, the boys and girls of Camp Zip Fire recited the Pledge of Allegiance. This part of the daily routine was always one of the Sassafrases' favorite moments.

Next, Chaz stepped forward and turned to look back at the boys of his cabin. "Okay, Mustangs. Today is our day to lead out in the Camp Zip Fire motto. Are you ready?"

Affirmative nods came from every member of Blaine's tribe. Then, as one, in loud clear voices, they recited the camp's motto.

Always do what's right
Even when it's hard
Against complacency
Be ever on guard.
Have courage like Zip
To face the fire
More than the minimum
Aim much higher.

A single breath after the Mustangs finished, all the other tribes repeated the motto in unison. Two breaths after the completion of the collaborative recitation of the motto, an excited murmuring began rumbling its way through the crowd of campers. Time for breakfast!

The swarm of campers were supposed to walk in an orderly fashion from the flagpole to the dining hall, but it hardly ever happened that way. As Blaine run-walked with the clump of his best buddies toward sustenance, they started talking about what was on their minds.

"We've still got the record for the fastest time on the old obstacle course," Jamal said with vigor. "But bro, the Sorraia tribe is looking strong this year."

"They sure are," Jarod agreed. "Their two best athletes looked good on the volleyball and basketball courts. I bet they'll do good on the obstacle course, too."

"But do you really think they can beat our record?" Blaine asked.

The group of boys around Blaine shrugged.

"We can probably keep the record on the old course, especially now that you're back. And, Bro, what about the new course down at the waterfront?" Matt exclaimed.

"It looks totally sick! When do you think we'll get to try that thing out?" Blaine asked.

"Tomorrow," an adult voice said from behind them. They turned to see Chaz Throckmorton, smiling.

"Aww, man, we have to wait until tomorrow?" Davy winced. "Does that mean we have to zip around on the old course first?"

"That means you GET to zip around on the old course first," Throckmorton countered. "Zipping around on the old obstacle course is nothing to be bummed about, fellas," the counselor continued as they reached the dining hall and the long line outside of it. "You guys love that course with all of its zip lines, crazy obstacles, and cool science."

"Oh, no, here we go again," Jamal said, rolling his eyes. "Chaz just said, 'Cool science.' You know that means he is going to say more."

Jarod, Matt, and Davy rolled their eyes too and sighed at the prospect of their camp counselor broaching this evidently undesirable subject.

Blaine looked at his friends in wonder. Back before he had launched off into this summer of adventure and learning, he used to have the same attitude toward science. He sure had changed, but he wasn't sure he wanted his friends to know this quite yet. What would they think about him if they found out he was now a big fan of science?

"Just like the new course will, the old course can teach us a lot about science and physics," Chaz said to the boys. "For example, gravity,"

"Gravity, wooka wooka wooka gravity!" Blaine suddenly blurted out.

His four friends looked at him with eyebrows hitting hairlines.

"What in the world was that?" Jarod asked, laughing with the rest of the boys.

The Sassafras boy simply replied with a big, nervous, toothy grin.

Chaz grinned, entertained by Blaine's strange outburst. "Gravity is an invisible force that pulls two objects together. It was one of the first fundamental forces of life to be discovered. The force of a gravitational pull depends upon the distance between the two objects and the mass of the two objects," the camp counselor explained. "The closer the objects are, the greater the force of gravity between them and the objects that have a greater mass produce a larger gravitational force."

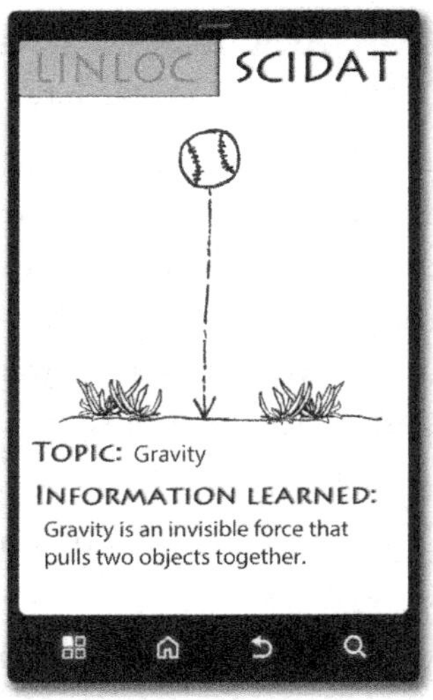

Blaine nodded while his friends rolled their eyes.

"The area in which the effect of gravity is felt is called the gravitational field, the center of gravity is the point at which the weight of an object appears to be located."

Even as Chaz continued to scientifically expound, Blaine began to vaguely remember his counselor talking like this before. He was sure this wasn't the first time Throckmorton had shared knowledge in the form of scientific subjects. It's just that Blaine had dismissed it in the past and remained uninterested.

No more.

Now Blaine was all in. However, he wasn't quite ready for his brothers from other mothers to know about this fresh interest. He'd keep quiet about it for now.

"Mass is the amount of matter something contains. It never varies. And weight is a measure of the pull of gravity on an object's mass. For example, this is why we would weigh less on the moon than we do on Earth and also why we would weigh more on Jupiter than we do on earth," Chaz continued as they shuffled forward, now inside the dining hall and getting closer to the long row of stainless-steel countertops where breakfast was being served.

"Gravity is the force that keeps the planets in orbit around the sun. Also, gravity is the force responsible for creating tides in the ocean. This is how that works: the gravity of the moon pulls on the ocean on the side of the earth that is closest to the moon, causing it to rise. As the two bodies rotate around each other, the strength of the force changes, causing tides. Isn't that cool?"

Blaine accidentally nodded, but his friends didn't.

"C'mon, Chaz," Matt laughed. "Not only is that not cool. What in the world does it have to do with the old obstacle course?"

Throckmorton returned his campers' laughs and partially conceded. "Okay, all right, I went on a bit of a roll there, but gravity totally pertains to the obstacle course! The reason the zip lines zip down, and the reason so many other obstacles are difficult and fun to navigate, is all because of gravity!"

Blank looks appeared on the faces of Jamal, Jarod, Matt, and Davy.

"Okay, okay, that's enough for now," Chaz laughed. "We're at the front of the line, so pay attention and shape up because you know how Grizz can be."

Friction Frenzy

Tracey gulped. Grizz's rules for the meal lines always made her a little nervous. The enormous man was an excellent cook. Camp Zip Fire's food was way better than the food at most camps, but he was REALLY particular about the receiving of that food. First, she did not know how a man of his considerable girth could get from the flagpole to the dining hall so quickly. On top of that, he could somehow change from his staff uniform to his 'chef whites' in a flash. But none of that made her nervous.

What made her nervous was the strict guidelines campers were supposed to follow in order to get their vittles, as Grizz called them. There was to be no pushing, no shoving, no cutting in line. When you reached the front of the line, you were to grab one tray, one plate, one bowl, one fork, one spoon, and one butter knife, even if you didn't plan on using all these items. Campers were to move to the right, sliding their trays along in front of them, keeping an ample space between themselves. Once a camper got to where Grizz was standing, that meant he or she had finally reached the vittles. The only thing now separating the camper from a hot meal was a sneeze guard and the chef.

The boys and girls could look Grizz in the eyes, but they weren't supposed to talk to him. They were to point with their entire hand, palm up, at the food they wanted and skip the items they didn't want. No grossed out faces were allowed. If Grizz filled the plate and bowl, it meant they had done things correctly and were free to move on to the cups and drinks section of the line, which was much less stressful. If, for some reason, Grizz did not fill a plate and bowl, the camper had to go to the back of the line and start again.

No whining. No questions. No grossed-out faces.

Every meal there were kids who didn't do it right. They had to start again from the back of the line. Tracey did not want to be one of those kids.

The Sassafras girl glanced up ahead and saw that Blaine's tribe, the Mustangs, were already at the drink station, meaning they'd made it past Grizz without incident. She hoped her tribe could do the same.

The Appaloosa tribe was normally pretty good in the vittles line, but before they reached Grizz, Marci and DeeDee said something about the zip lines on the old obstacle course. This got their counselor started sharing about science. Tracey was interested by what Hattie was saying about friction. She was just worried all this talking was going to get them in trouble with Grizz.

"Friction is a force that opposes the motion of an object as it passes another. This force pulls against the direction of the movement," Cates said, being expressive with her hands as she talked. "Another thing friction does is it changes kinetic energy into heat energy, causing objects to heat up. The rougher the surface, the stronger the force of friction. The harder the objects press against each other, the stronger the force of friction."

Tracey had never remembered Hattie being this passionate or knowledgeable about science. She must have missed that fact in the past. It really impressed Tracey right now, but not her friends. Joanna, Janna, Marci, and DeeDee looked more than a little uninterested. Tracey figured that used to be her posture as well. That, of course, was not her posture anymore. But for now, she wanted to fall in with her friends and pretend science was of no interest to her.

"Without friction, there are a lot of things we couldn't do," Hattie Cates continued, "like walking! We couldn't walk because our shoes could not grip the ground without friction. We also couldn't grab anything without friction because our fingers wouldn't be able to hold the objects."

Oh no, Tracey internally jolted. They had reached the stainless-steel counters. It was time to use some of that friction Hattie had just talked about. They needed to grab and grip some trays, and they needed to stop talking to ensure they didn't upset the chef. But instead, the counselor just went right on sharing.

"Friction can be both positive and negative. For instance, it is positive when it is the force responsible for slowing a vehicle down when we hit the brakes. But it is negative when it does things like cause wear and tear on machines."

Every girl in the Appaloosa tribe now had their trays, their plates, their bowls, and their silverware. They were all sliding to their right. They would be in front of Grizz in a matter of seconds, but still, Hattie shared on.

"However, we can reduce the friction different machines and objects experience by using materials like oil, grease, and other lubricants. We can also use ball bearings, which reduce friction by causing objects to roll over each other instead of dragging over each other."

Tracey froze, except for making another gulp. Their counselor was the one leading their tribe in the vittles line, and Cates now stood directly in front of Grizz. But she was not deciding what food she wanted. She was not pointing with open palms. She had forgotten to grab both a bowl and a spoon, and she was standing still, looking back at her campers, talking science and holding up the line.

"Oh, yeah, I wanted to also say something about air resistance," Hattie continued, seemingly oblivious to the chef's ire. "Air resistance is the friction that occurs between air and an object moving through it. That's why we give planes, vehicles, and other objects streamlined designs to reduce the friction, allowing those objects to move more easily."

Tracey probably would've thought it was pretty funny that Hattie had just said move more easily if she wasn't so scared about Grizz's impending response to Hattie not 'moving more easily' and not following at least half his rules.

All at once, the camp counselor realized all her girls were standing statue-like. She looked confused for a second and then realized where she was in line. She turned toward the big bearded chef with a friendly smile. "Oh, hey, Grizz."

Grizz responded with neither frown nor smile, exhale nor inhale, words nor motion. He stood still, looking at the cheerful camp counselor. Then, finally, after what seemed like way too long and without breaking eye contact with her, the burly cook filled Hattie's plate with substantial piles of scrambled eggs, biscuits with gravy, and bacon to garnish.

"Thanks, buddy!" Cates exclaimed, totally unfazed. Apparently, counselors got a pass from Grizz's strict line etiquette rules. Tracey shrugged, quickly resetting her posture. She didn't want to take any chances with Grizz sending her to the back of the line.

Once the whole Appaloosa tribe was safely seated far away from the vittles line, Tracey watched the second girl from the following tribe get sent to the back of the line. She sighed, feeling bad for the girl. As a distraction, Tracey turned and asked her counselor, "Hattie, what does friction have to do with the zip lines?"

"Oh yes, that's why I originally started talking about friction, right?"

Tracey nodded. The others rolled their eyes. "Oh no," Tracey thought. "Have I shown too much interest? Will Joanna, Janna, Marci, and DeeDee still want to be friends with me?"

"Well, I'm probably biased," Cates said, unaware of Tracey's internal worries. "But I think we have the best and the fastest zip lines around. We not only keep our lines cleaned and polished, but we also use the highest quality trolleys, wheels, pulleys, and bearings to reduce as much friction on our zip lines as possible."

"Speaking of trolleys, wheels, pulleys, and bearings," a male voice said behind the girls. "Let's get out on those zip lines right now!"

It was Chaz Throckmorton, and he and his Mustang tribe had scarfed down all their food and were already clearing their trays.

"Wait a second, Chaz, I thought you guys were headed to horseback riding first thing this morning and then later to the old zip-line course?" Hattie asked in a good-natured way that made it clear they were friends.

Throckmorton paused with a momentary quizzical look on his face. "You know what, Hattie, you're absolutely right. It's your tribe that's up first on the old obstacle course this morning." He laughed. "All right, boys, you heard the woman. Off we go to the horse barn!"

Blaine gave another small wave to his sister and then went with his tribe mates to put their trays, dishes, silverware, and trash away, being careful to follow all of Grizz's particular discard rules. Once they were trayless and free, they bounded out to the horse barn.

Blaine's favorite part of Camp Zip Fire was the obstacle course and the zip lines, and he was really excited about trying out the new obstacle course down at the waterfront, but he was also a big fan of the other activities that Camp Zip Fire had to offer, including horseback riding. Historically, Camp Zip Fire had been founded and built around Montana's horse and ranch culture.

The Mustang tribe went from the dining hall, passed the flagpole, and back by their cabin. They passed the basketball and volleyball courts, ending at the big horse barn and riding pasture. There were about twenty boys in the Mustang tribe, and Blaine knew most of them. They had all been a part of this tribe for several summers together. He considered Jamal, Jarod, Matt, and Davy his best friends. But as they arrived at the barn, Blaine noticed a kid he didn't know at all—a bashful-looking kid that didn't seem to interact with anyone. He was short, super skinny, and seemed like he might be a couple of years too young to be in this tribe.

Blaine nudged Jamal, made a head motion toward the kid and asked, "Who's that?"

Jamal answered in a whisper that seemed none-too-pleased. "The kid's name is Timmy Brown. None of us know where he came from. He showed up the day before you did. But bro, the new obstacle course tomorrow, you know? It's a competition, and it's a team effort. How are we going to beat the Sorraia tribe or anyone else with a scrawny kid like that on our team?"

The tribes were named after different horse breeds. Tracey liked that, and Tracey liked her tribe's name the best: The Appaloosa tribe. However, the jury was still out on the Jennet tribe. She didn't know if it was the name of the tribe or the girls that were in the tribe. It was probably the latter, but Tracey wasn't really a big fan.

The Appaloosa tribe had the record for the old obstacle course, and she was hoping they could set the inaugural record on the new course tomorrow, if for no other reason than to beat those girls from the Jennet tribe. Some girls from that tribe were especially pestering. Nothing serious—most of the back-and-forth between her tribe and theirs was just amiable competition, but Tracey really wanted to beat them on both obstacle courses and so did the rest of her tribe mates.

Joanna, Janna, Marci, and DeeDee were unquestionably Tracey's BFFs. They had all been friends for what seemed like forever. The Sassafras girl loved all her other dozen tribe members as well. They all got along and worked, played, and competed well together. But even now, as the Appaloosa tribe left the dining hall and headed over to the old obstacle course, Tracey was noticing a girl with them she didn't recognize. She was taller than all the other girls and thin. She also seemed a little standoffish, and she wasn't interacting with anyone else.

Tracey sidled up next to Joanna as they walked. "Have you met the new girl?" She asked quietly.

Joanna shook her head. "Nope, but I heard Hattie call her Gretta. Gretta Brown, I think. She showed up the day before you did. I think her brother is in your brother's tribe."

Janna, who'd heard her friends whispering, came alongside and joined the hushed conversation. "Her being here makes me a little worried. I mean, I don't want to be mean, but the obstacle course is a team effort and basically a speed competition, and I mean, just look at her. She's like a string bean with no muscle on her bones. Do you think she can really help us win?"

CHAPTER 5: ZIPPING ACROSS THE BIG SKY

Balance Beams

Even now, as she stood at the edge of the platform, ready to jump out on the first zip line, Tracey played through the story of the campground's founder and his horse in her mind. She wanted to be bold like Biff. She wanted to be courageous like Zip. Especially if it meant helping the Appaloosa tribe keep the speed record on this old obstacle course. The story of the camp's founder rolled through her mind.

The fire burned out of control faster than Biff Clark expected. He was trapped. Lightning had struck an old dead lodgepole pine, which had acted like a lit matchstick among the tall carpet of dried out sagebrush. A wall of fire encircled his ranch house, and the flames grew higher. Luckily, his wife and children were over in Billings visiting her parents, so this disaster was of no threat to them, but it looked like this very well could be the end for him.

Biff had put his entire adult life into developing this ranch. Was it going to be burned to ashes? Would the fire take this house he'd built with his own two hands? Would it take his barns and silos? Would it spread out and scorch his 300 acres? Would it take his life and the life of his horse? Biff Clark stepped in the stirrup with his left foot and kicked his right foot up and over to mount his trusty steed now. The rancher needed a higher vantage point, but the wall of fire was still a little higher than his line of sight.

Biff patted his horse on the neck and spoke into his old friend's ear. "This is not looking good, Zip. Not good at all."

The Holstein tilted his back toward his rider in acknowledgement and then snorted as if he understood. Zip then faced the fire. Another snort burst forth from the horse's nostrils. A resolute snort of determination. His horse was in no way ready to concede to the prospect of them being goners.

Zip raised up on his hind legs and let out a loud and courageous neigh. All at once, Biff Clark knew what was about to happen. His tried-and-true Holstein horse was about to make an attempt to jump through the encasing wall of fire! The rancher was tempted to try to stop the animal because he knew the wall of fire was wide and far too high. It would be virtually impossible to make it through alive.

But what other choice did he and Zip have?

It immediately became apparent it wasn't actually up to the rancher at all. Zip was going to go for it with or without him. The big Holstein reared back on his hind legs once more and went for it. Biff clung to the reins as Zip charged forward, straight toward the fire.

Every muscle in the horse's body was surging, his mane was wildly waving, and his snorts were coming in quick rhythm and utter determination. Then it was time to jump. With no hesitation, Zip left the earth for the sky.

It all happened in slow motion. Heat lashed out with fury, attempting to burn. Tongues of fire licked up at them, attempting to scorch. But Zip had aimed much higher. To the rancher's shock and amazement, his brave Holstein hadn't just made it through the fire— Zip had cleared the burning wall by a good foot. Zip's hooves came crashing down, clean and clear on the other side of the fire in perfect succession, with his rider still safely on his back. They had escaped the circle of dissolution! They had jumped the wall of fire!

Now, on the other side of the flames, Biff had access to creek water, which he painstakingly used over the next several hours to put out the flames. In the end, his ranch house was damaged but saved. His barns and silos were not touched, and the vast majority of his 300 acres of Montana ranch land was left unscathed. Most importantly, his life had been spared, and it was all because of the courage of his trusty steed, Zip.

Tracey silently quoted the camp motto as she awaited her turn. That always helped her stay focused in the competitions here at Camp Zip Fire.

"Always do what's right, even when it's hard," the Sassafras girl repeated in her mind. "Against complacency, be ever on guard. Have courage like Zip to face the fire, more than the minimum, aim much higher."

"You're not repeating the camp motto in your mind in order to focus, are you?" Tracey heard a giggling voice to her left ask.

The Sassafras girl's face looked over and saw a girl from the Jennet tribe standing a few feet away, also preparing to soar out on the first zip line. Tracey had known she was here, of course, but she had kind of just gotten lost in her own thoughts. Instead of answering with words, Tracey responded with an attempt at a confident sounding laugh, which came out sounding a little more like a gurgling cough.

The Jennet girl's non-cough-sounding laughter continued as Tracey again turned her attention to the course in front of her. The way it worked was that the two tribes raced the course side by side. They started on a tall wooden platform, harnessed and clipped into zip lines. The moment they were given the signal by the counselor stationed there on the platform, they could launch out onto the line. It was a long and fast zip line that whisked them over treetops, a stream, and a pathway that often held onlooking campers from other tribes.

They sailed over the distance to another platform, where another two counselors were stationed. The trained counselors would unclip them from the first line and guide them to the second line, which did not require a clip in. Campers held on as they zipped, but if they fell, they would land on a huge section of stretched netting. Fun, but it slowed the race time down. It was much better time-wise to zip to the end of the line and then drop on the netting.

After that, they climbed from the netting to another platform, which led to a bouncy wood-planked bridge. Across the bridge, they went to a set of rope swings. They flung themselves out on the rope swing and dropped onto a mound of spongy shredded pool noodles. From the shredded pool noodles, they moved onto solid ground where there was an obstacle called the Tire Jump. It was a line of ten tires buried halfway in the ground, about three to four feet from each other. They jumped from tire to tire and couldn't let any part of their bodies touch the ground. If they did, they had to start back at the first tire.

Last but not least was the Hamster Roll, where they got situated inside a human-sized version of a cushioned hamster ball. A counselor would then push them down a small hill toward a section of hay bales. Once a hamster ball stopped at the bales, they crawled out and in their dizzy state, they raced over the finish line.

It was a crazy obstacle course. One that the Appaloosa tribe had always been good at. Tracey hoped this remained the case today. She would be the first girl to race a Jennet girl for her tribe. The girls behind them could start the course when Tracey and the other girl got to the wood-planked bridge. Once all the members of the tribe passed over the finish line, the time was stopped. Only two tribes raced at a time, but every tribe present would have time to compete. The fastest time won.

Tracey waited for a signal from the counselor here on the first platform. She took a deep breath and quoted the camp motto one more time in her mind. Then she saw it: the signal from the camp counselor. Tracey jumped out onto the zip line for all she was worth!

The Mustang tribe had moved from horseback riding to leather, working, and then to lunch. After lunch, they had attacked the old obstacle course—or had been attacked by it was probably more like it. The whole time they had moved from activity to activity, Timmy Brown had moved by himself. The only person who had talked with him was their counselor, Chaz Throckmorton. None of the perennial Mustang campers had had any kind of conversation with him or asked him questions. Blaine felt really badly about that, but he was trying to stay focused on the major tasks at hand.

They still had the overall record for time, but the day out at the classic Camp Zip Fire obstacle course had gone pretty badly for the Mustang tribe. Not only had the Sorraia tribe beaten their time today on the old course, but so had the other two tribes. If today was a warm up for the new waterfront obstacle course tomorrow, things weren't looking good for Blaine and his friends. It was difficult to tell whose fault it had truly been. The entire Mustang tribe had raced slowly and sloppily today, but the blame was currently being directed at the new kid, Timmy Brown. Blaine hadn't noticed Timmy making any mistakes worse than any others, but his fellow tribe mates were making it sound like it was all the little guy's fault.

"I mean, did you see how quickly he fell off the second zip line?" Davy said.

"Yeah, and how bad he did at the tire jump?" Jamal added, with no mention of how he had fallen off the second zip line, too.

"We never had a real chance to win today with him on our team," Jarod concluded, ignoring his own mistakes on the course.

Blaine didn't agree with his friends, but he kept quiet. He didn't want to say anything that might expose how poorly he'd done on the course. But he was also afraid they were talking loud enough for Timmy to overhear them.

"Whatever, fellas," Matt said with slaps on all his friends' backs. "Today was a wash, but that doesn't really matter. We still hold the all-time record on this old thing."

"Yeah, bro, and what's really important is how we do on the new obstacle course tomorrow. That's where we need to send it," Jamal added with gusto.

"Yeah, but how are we going to do that with him on our team?" Davy said, pointing at Brown.

Blaine was certain this time Timmy had heard them and that the short skinny guy had seen Davy point. But instead of going over to apologize like he knew he should, Blaine chuckled along with all of his buddies.

"Okay, Mustangs," Chaz said, approaching. "It's now time to head down to the waterfront!"

"To try out the new obstacle course?" Davy shouted.

"Nope," Chaz said. "Time to go canoeing."

A collective sigh released from Throckmorton's campers.

"Oh, c'mon, Mustangs," Chaz cheered. "You guys love

canoeing. Anyhow, you'll get your first look at the new ropes course today, but you won't get your first swing at it until tomorrow morning after breakfast."

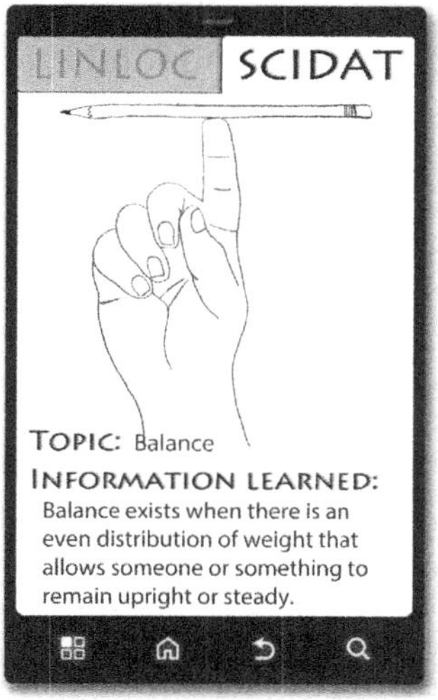

"Balance exists when there is an even distribution of weight that allows someone or something to remain upright or steady," Hattie Cates said as she switched her paddle from the left side of the slender boat to the right. "Every object has forces, or pushes and pulls, that act up on it. These forces can cause an object to move in some way or they can cancel each other out, causing the object to remain at rest. An object is in balance when all the forces that push or pull on it have caused it to remain still."

Tracey, who shared a canoe with her counselor, had asked the question about balance. She currently looked up from her spot in the boat at several high balance beams that spanned a long distance above them. The Appaloosa tribe was paddling around the small lake trying to take in the grandeur of this new Camp Zip Fire waterfront obstacle course they were seeing for the first time.

"A camper walking across one of those balance beams will have the force of gravity pulling them down toward the water, and at the same time the force of the balance beam holding them up. Or another example of balance would be like a book sitting on a table," Cates said as Tracey continued to follow along with the science her counselor shared. She hoped her friends were far enough away so as not to witness her excited interest in the information.

"The force of gravity would pull the book down toward the floor, and the force of the table strength would hold it up. The book would remain on the table unless the table was tilted causing the force of gravity to become greater, making the book fall to the floor."

"Which would be like someone leaning too far to either side of one of those balance beams, right?" Tracey asked. "Even a slight lean could increase the force of gravity, causing them to tumble off the beam down into the lake."

"Exactly!" Hattie Cates confirmed. "Wow, Tracey, you seem like a different girl this summer! You never showed this much interest in the scientific side of things before. What have you been up to before you got here?"

The Sassafras girl smiled sheepishly, hoping none of the other campers had heard her and Hattie talking. The way-up-high balance beams were just a small part of this new course. This thing had to be at least twice as big as the older land-bound course they had completed this morning.

Tracey could clearly see the big floating blob-looking thing she and Blaine had landed on last night and the large multi-storied dock behind it. Branching out from the big combination wooden-framed and steel-framed structure were zip lines, slides, wall climbs, swinging rings, the blob, and other floating obstacles. Plus, the balance beams Hattie and Tracey had just been talking about. The Sassafras girl was sure this huge thing was going to end up being fun, but right now, it just looked intimidating.

Suddenly, a shrill high-pitched noise cut through both the air and Tracey's wonderment.

It was the lifeguard, standing up on his elevated chair, blowing his whistle and pointing. The lifeguard was none other than big Grizz himself, now adorned in red shorts, a white tank top, and lots of sunscreen instead of his staff uniform or his chef whites. He was pointing at the canoe that DeeDee and Joanna were in.

Evidently, they had just used their paddles to splash another canoe, which was against Grizz's waterfront rules.

Tracey could see that both of her friends had been splashing the boat with Gretta Brown and another camper. Gretta was dripping wet. However, the other girl was not. Had Tracey's friends just targeted the new girl, or was it just a playful accident? Tracey couldn't tell, but it looked like Joanna and DeeDee were currently less concerned about apologizing than they were about giggling.

This morning, during the day's first camp activity, the Appaloosa tribe had done okay on the classic obstacle course. They were not good enough to set a record . . . or to beat the Jennet tribe. The blame for the lackluster performance was placed squarely on the shoulder of Gretta Brown just because she stumbled a little on the bouncy wood-plank bridge. This wasn't fair at all, in Tracey's mind. All of them had made tiny mistakes, too. Mistakes that added up.

However, this hidden sentiment had not yet moved Tracey to say anything about it to her friends. The Sassafras girl had remained quiet as Joanna, Janna, Marci, DeeDee, and the others had cast shade at Gretta behind her back. Tracey made a mental note to bring it up later, but for now she would just focus on maintaining her current friendships and staying in the right head space to rock the new obstacle course tomorrow.

Blob Buoyancy

The Mustang tribe passed the Appaloosa tribe on the trail as they both headed to their last activities of the day before dinner. The girls were headed to horseback riding and the boys to canoeing.

"Did you get to see the new obstacle course down at the waterfront?" Blaine asked Tracey as they passed. "Is it cool?"

"Yes, to both," Tracey answered, walking by in a clump with her BFFs.

"Sick," the Sassafras boy exclaimed to Jamal, Jarod, Matt,

and Davy, who were walking with him.

Being led by Chaz, the Mustang tribe hustled, but didn't run because that was against Grizz's rules. When they reached the waterfront, the boys statue-stopped in awe. This new obstacle course was outstanding!

"All right, boys," Throckmorton said. "We are here for the canoes, but since I know you are all wondering, I'll tell you how this obstacle course works."

All his campers were listening, but none of them moved a muscle as they were all still too astonished to respond.

"Just like on the classic course, fastest time wins. Two tribes will race at a time; every tribe gets a turn. On this bad boy, you start at the very top of that immense structure. And be warned, fellas, you are never harnessed or clipped in here so you can fall at pretty much any time. Yes, you will fall into the water, and yes, you will have life jackets on, but that can slap, and not in a good way." Chaz paused as he even now grabbed the life jackets from a nearby rack for the canoes.

"Also, just like on the classic course, your first obstacle is a zip line," he said, tossing the life jackets to his campers. None of them were catching the jackets because they were all standing and staring at the gilt-edged obstacle course.

"You jump off the top level next to your competitor and zip over the water toward that very colorful and well-padded tower you can see there," Chaz continued, now pointing. "From that platform, you must make your way over a series of rather long balance beams. After the beams, you swing across half a dozen hanging rings. Then you careen down those long slides, splash into the lake, grab your bearings, and then swim the short distance to those big padded floating circles, which are referred to as 'lily pads.' Maneuver your way across the row of tipsy and precarious lily pads back toward the multi-storied platform where you will encounter a wall climb. Climb back to the top of the platform where a stationed counselor will guide you to the side opposite the zip line. From that side, you will jump down onto that huge, inflated blob."

For some reason, the word 'blob' seemed to shake most of the boys from their awestruck state into low grumbles of acknowledgement.

Throckmorton continued, "Once you land on the blob, carefully crawl down to the end and situate yourself in the marked area. Wait there until the tribesman who is racing behind you jumps from the platform down onto the other end of the blob, launching you high into the air."

Excited oohs and aahs echoed from the Mustangs.

"Now if you'll look out in the water beyond the blob, you can see three large floating numbered rings: one, two, and three. One being closest to the blob, three being the furthest," the almost equally as excited counselor explained. "For every camper that lands inside one of those rings, time is shaved off the team's race time: five seconds off for ring one, fifteen seconds for ring two, and thirty seconds for ring three!"

More oohs and ahhs erupted from Blaine and his tribe members.

"Post blob launch, you will then swim as fast as you can past the finish line." Finished with his explanation and description, Chaz now joined his campers in a long moment of staring at the promise of recreation.

"Whelp, that's it, boys: the new obstacle course," the counselor finally said. "But we'll have to wait until tomorrow to tackle it. Today we're here for the canoes."

At this, most of the Mustangs slipped their life jackets on and headed down to the water's edge to a group of waiting canoes. Blaine lingered in his spot, trying to take in this new amazing course.

"It floats," he said in awe.

"Because it's buoyant," a voice said. "As are the canoes, which we are supposed to be getting into right now."

It was Chaz, and with a smile, he was directing Blaine to the boats, which the Sassafras boy was fine with as he enjoyed canoeing. Blaine heeded his camp counselor and went down to climb into a boat. He was the last of the Mustangs to do so. Therefore, he got to share a canoe with a counselor.

"Buoyant, huh?" Blaine asked as they pushed away from shore.

"Yep," Throckmorton confirmed. "Buoyancy describes how an object will float or not on water and even in the air. An object will float on water when the upward force of the fluid matches the downward force of gravity. But if the downward force is stronger than the upward force, the object will sink."

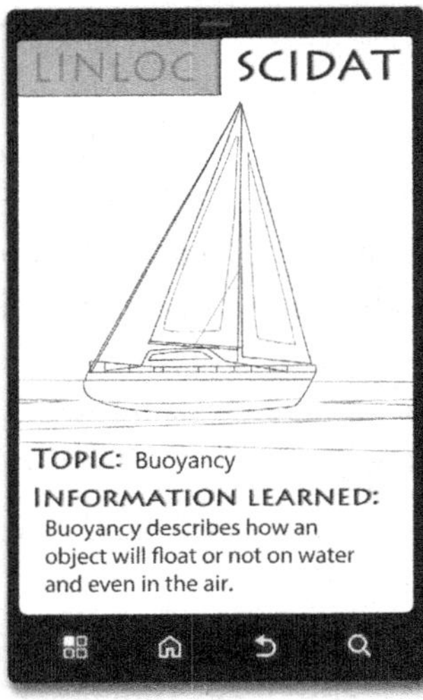

Blaine took a glance over toward the lifeguard station and saw they were all boating under the watchful eye of Grizz. He sure hoped all his friends resisted the urge to horseplay on the lake. It seemed like at least one canoe accidentally tipped over every time they were out here.

"And just to clarify," Chaz said from the back of the canoe, "The downward force results from gravity on the mass, otherwise known as weight. The upward force is called 'upthrust.' It depends on the weight of the object and the volume of the fluid that is displaced or pushed aside. This is caused by the pressure of the fluid on the object."

Uh oh, there was some excessive-looking splashing happening out further in the lake.

Grizz stood up, whistle in mouth, blowing and pointing at even the slightest infraction.

"The Archimedes Principle says that the upthrust acting on an object is equal to the weight of the fluid that the object displaces. Does that make sense?" Throckmorton asked, still sharing information.

"It sure does," Blaine answered loudly enough for his camp counselor to hear, but hopefully not loudly enough for any of his tribe mates to hear. He didn't really want any of them to know he was this into science.

"All that leads me to mention density," Chaz said, more than likely unaware of Blaine's attempt at social preservation. "Density is the measure of an object's mass compared to its volume or how much matter it contains, compared to how much space that matter takes up. An object will float in water if its density is less than the density of water. The heavier and more packed together an object is, the more likely it is to sink. This is why stones tend to sink and corks tend to float."

Blaine nodded in understanding as his counselor and local expert continued. "Submarines have something called a ballast that can be filled or emptied, allowing them to either sink or float. Fish have a swim bladder that helps them do the same thing."

"Okay, so this blob thing," the Sassafras boy said, pointing. "It floats on top of the water because it's filled with air, right? Almost like a fish bladder or an empty ballast tank or whatever that thing in a submarine was?"

"Exactly," Chaz confirmed. "Man, Blaine, I've never known you to be this interested in science," Throckmorton said with a chuckle.

Blaine returned the laughter, then said, "And I never remember you talking this scientifically about things."

"You must've just missed it," Throckmorton said, still

laughing.

A series of three long whistles suddenly pierced the late afternoon sky, "Whelp, there it is," the counselor sighed. "The obligatory canoe tip over."

Chaz was correct—a canoe had gone bottom over top. As the two from the boat quickly resurfaced thanks to their life jackets, Blaine saw a wet mess of curly hair and a skinny little soaked head of hair. It looked like the tipped canoe had held Jarod and Timmy. Jarod was laughing, along with all the boys in the surrounding canoes.

Timmy, however, was not laughing. Blaine couldn't tell if Brown was scared, angry, embarrassed, or all three.

Had Jarod tipped the canoe on purpose just to fluster the new kid? Had someone else done it? Blaine knew he needed to say something to his friends about how they were treating Timmy. He was sure he would get around to doing just that . . . maybe . . . you know . . . sometime later.

The soft orange glow on everyone's faces from reflected dancing flames, the warmth the campfire offered in the Montana evenings, and the close proximity of known and cherished peers usually made Tracey feel super cozy but not tonight. She'd been able to take part in many of her favorite camp activities at her favorite camp in the entire world. She loved the invisible zip lines and all the new places they'd taken her and her brother this summer, but it sure had been nice to immerse herself in someplace familiar, just for the day.

Right now, Tracey should feel happy. Happy and warm here at the campfire. But she didn't. Tonight she felt cold. Why did she feel cold? Maybe it wasn't her body that was cold; maybe it was her heart. Why did her heart feel cold? Maybe it was because she had refused to do what was right.

After horseback riding, her Appaloosa tribe had tromped back to the dining hall for dinner. No one had been sent to the back of the line. They were too tired and hungry to do anything other than stand quietly in line. They'd gotten their bellies full and then they'd headed here to the fire pit. It was in a clearing down from the flagpole. This was the last stop of the day for every tribe and every camper.

Soon a camp counselor would stand up and start telling a campfire story—a story that usually had something to do with a person showing courage or valor in the face of some kind of adversity. It usually was a story that would be like Camp Zip Fire's foundational story of Biff and Zip surviving the fire. Tracey usually looked forward to this last activity of the day, this story time around the campfire, but not tonight.

Tonight she didn't want to hear any stories of courage or valor. These kinds of stories would just remind her how she had shown neither courage nor valor. Basically, from the Appaloosa tribe's first activity to the last, Tracey's friends had been making fun of Gretta Brown. Tracey had said nothing and had done nothing to stop them.

Always do what's right
Even when it's hard
Against complacency
Be ever on guard
Have courage like Zip
To face the fire
More than the minimum
Aim much higher.

Blaine rolled through the camp motto in his mind as he

stared at the fluttering flames and listened to the counselor telling some story about some hero choosing to do the right thing even when the odds were stacked against them.

The Sassafras boy sighed and let himself slouch down even further. "I'm nothing like this hero. I'm not like Zip or Biff either. Really, I'm more like the Man With No Eyebrows than I am like any kind of hero. I'm a villain," Blaine thought. If he was going to refuse to do what was right. If he was going to be complacent, if he was going to choose fear instead of courage, and if he was going to do less than the minimum, then that had to mean he was a villain, a bad guy, right? All day, the members of the Mustang tribe had been belittling the new kid, Timmy Brown. And Blaine had done nothing. He'd been a coward, not a hero.

The counselor who'd been sharing finished the story and took a seat. Another counselor then stood up and gave the nightly invitation.

"Okay, now it is time for any campers who want to briefly share."

This announcement received the usual hush from the large group of campers. In all the years the Sassafrases had been going here, only one or two kids had ever been brave enough to stand up and share. Usually, everybody was ready to get back to their cabins for a good night's sleep, which was often preceded by lots of talking, joking, giggling, horseplay, and pillow fights.

Tonight was the same. There was no response from any of the boys or girls. No camper had anything they wanted to share.

Suddenly, Blaine stood up. What was he doing? Jamal and Jarod had looks of shock on their faces. Matt and Davy looked like they'd swallowed frogs. Counselor Chaz looked confident and proud.

"I have something to share," Blaine heard himself say.

Every camper and every counselor was giving the boy their

full attention. "I've made a lot of mistakes today," Blaine started.

Many looks of confusion could be seen on the sea of orange-tinted faces.

Blaine found Timmy Brown's face in the crowd and made eye contact with him. "Timmy, I'm sorry man. You've been getting blamed and talked about all day, and instead of sticking up for you, I either remained silent or even kind of joined in. I was wrong about that. I should've befriended you and officially invited you to be a part of the Mustangs. I've been wrong, bro. I'm sorry."

If silence could reverberate, it did in this moment. Even the surrounding sounds of Montana nature seemed reposed. Blaine couldn't really believe he did what he'd just done or said, and certainly didn't know what to do now.

Suddenly, he realized someone standing next to him. It was his twin sister.

"I've been wrong, too," Tracey said as she located Gretta Brown among the campers. She looked at the girl with regret and empathy rolled into one expression. "Gretta, today you have been treated poorly, and what's even worse, I didn't stand up for you. I'm so sorry, Gretta."

The expressions of both the Brown siblings were virtually impossible to read. However, Joanna's and Janna's faces looked petrified. Marci's and DeeDee's expressions screamed panic. On the other hand, Counselor Hattie's face was exultant.

The apologies hung in the air—what was going to happen next? Had Blaine and Tracey just made complete fools of themselves? Had they isolated Timmy and Gretta even more?

All at once the stone-cold silence was broken by the sound of soft crying. It was Joanna. She let the floodgates open as she said, "I'm so sorry, Gretta! I'm wrong too!"

Janna, Marci, and DeeDee quickly joined Joanna. The four got up, joining Tracey up front. "We are so sorry, Gretta! All of us!

The entire Appaloosa tribe!" Marci wailed for the group.

"Bro, Mustang tribe!" Another voice suddenly called out. "Blaine is so right. The way we treated Timmy today is totally messed up. It's wrong."

It was Jamal. As he apologized, he got up to join Blaine. He was followed by Jarod, Matt, and Davy. All the boys made eye contact with Timmy Brown as they stood there, and they all looked like they were choking back emotion.

"Timmy, we were way off today bro. We're sorry!" Matt said for the group.

Timmy's face remained undecipherable, as did Gretta's. What were the two feeling?

CHAPTER 6: FLYING OUT OF IGUAZU

Energetic Falls

The next morning, Gretta Brown waltzed across the balance beams like an Olympic gymnast. Timmy Brown swung across the hanging rings like a superhero. The two new kids helped the Appaloosa tribe and the Mustang tribe dominate the new waterfront obstacle course, lighter from the forgiveness given and received the previous night. The two teams set the inaugural speed records for a girl's tribe and a boy's tribe, respectively.

No other tribes had come close to the times the Appaloosa tribe and the Mustang tribe had set. Even so, all the campers had fun on the new course. The children emitted lots of laughter and gasps of delight as they navigated the new and exciting obstacles the waterfront course had to offer.

The zip lines were low on friction and high on speed. The colorful padded platform was fun to land on and sometimes to slide and bounce off of. The beams and rings were scary, but fun. The slides were steep and fast. The lily pads were a ridiculous delight. The wall climb was a puzzle to solve. And the blob! However, the crowning moment of the whole obstacle course experience had been what one of the staff did. Grizz shocked everyone after the competition when he not only climbed down from his lifeguard chair and smiled, but he took that smile to the top of the multi-storied dock and announced that he would be the official blob launcher!

Blaine and Tracey had been launched high into the air several times by the huge camp staffer. All enjoyed every second of the flying and splashing as they played on the blob. The waterfront fun lasted for several hours into the summer afternoon, but as the activity lulled and the campers and counselors prepared to return to their cabins for a period of rest before dinner, the realization hit the Sassafras twins that it was time to leave.

Chaz and Hattie had shared all about gravity, friction, balance, and buoyancy. Now it was time to move on to their next learning location. Blaine and Tracey were, of course, having a blast zipping all over the globe, but they hadn't realized how much they had missed this place and their friends. It sure had been nice to be somewhere familiar for a couple of days.

It sure had been nice to be somewhere familiar, yet it wasn't home, and Summer was ready to be either home in Alaska or with the person she really wanted to be with—not Paul Sims and not even with Mrs. Pascapali, as sweet as she was. She wanted to be with Cecil Sassafras. Did Cecil not want to see her? Was Paul purposefully keeping her from Cecil? It kind of seemed like he was. But why?

She sauntered from the guest room where she was staying in Mrs. Pascapali's house out into the quaint living room. The hospitable widow was sitting in her favorite chair with knitting needles and a roll of yarn.

(Image #7: Summer and Mrs. Pascapali in the living room.)

"Well, hello there, Summer," Mrs. Pascapali said in her creaky-kind grandma voice. "Come sit with me and have a cup of chamomile tea."

The widow gestured to a pristine couch with a tasteful floral pattern design and a polished antique table where a steaming cup of tea was already waiting on a handmade doily. Summer genuinely liked and appreciated Mrs. Pascapali, but she would much rather be sitting in a messy basement lab drinking a lemonade with Cecil. Nevertheless, the female scientist sat down and said, "Thank you."

She reached for the small dainty cup of tea. As she did, she was surprised to see an envelope on the table addressed to her.

"Oh, what's this?" Summer asked.

"That was sitting in the mailbox for you, honey," the elderly woman said as she got back to her knitting.

Summer took a tiny sip of her tea, put her cup back down, and then picked up the envelope with a curious look on her face. Why was she receiving mail at Mrs. Pascapali's house?

The scientist opened the envelope and carefully pulled out what it held. It was a tastefully designed invitation that read:

> Dear Summer Beach,
>
> You are cordially invited to a junior high class reunion!
>
> Come back to the junior high school building and enjoy getting reacquainted with old friends and faces.

It then listed the date, the time, and the room number where the reunion would be held. The look on Summer's face went from curious to more curious and suddenly hopeful. A reunion! With former junior high classmates, including Cecil Sassafras!

A weird mix of joy and sorrow washed over the Sassafras Twins as they had zipped away from Camp Zip Fire in Montana to the country of Brazil, where LINLOC had told them they would study energy, kinetic and potential energy, heat, and changes of state, all with the help of a local expert named Augustus McShane. Their sight and strength had returned, and now they were trying to gather their bearings.

"Is this the heliquickter?" Blaine asked. "Trace, I think we landed in Summer's heliquickter!" The Sassafras boy referred to the flying vehicle their favorite local expert flew and operated. It was a machine that had been upgraded to travel twice as fast as a normal helicopter. Therefore, it was referred to as a heliquickter rather than a helicopter.

"It looks kind of like Summer's heliquickter," Tracey said. "But it's not. Just look how fancy this thing is."

Tracey was right. Inside this vehicle there were five plush, earth-toned leather seats, an updated interior that felt almost like a limousine, and a cockpit decked out with what looked like all the latest and greatest technology.

Suddenly, from somewhere outside the vehicle, Tracey and Blaine heard approaching footsteps. The Sassafrases leaned forward from their two leather seats to take a wider look out the open side door. At first, all they saw was the surrounding landscape—a lush, vibrant green forest that looked more tropical than coniferous. The sun looked as though it pondered its daily plummet. As they took in the beauty, the clues pointed to Brazil instead of Alaska.

An enormous giant of a man, who had to be close to seven feet tall, walked toward them. He wore what looked to the twins like a pilot's uniform, complete with shined shoes, dark slacks, and a white short-sleeve button-up that fit tightly on his massive muscular frame. The large man reached the open side door of the helicopter. When he saw Blaine and Tracey, he had a momentary look of confusion. One that quickly gave way to a look of acceptance as he shrugged his shoulders.

The man stepped up into the vehicle, making his way right past the twins into the double-seated cockpit. He took the helm without saying a word to the twelve-year-olds. His gargantuan frame made him look like a big tuna stuck in a tiny sardine can. Yet his confidence said he belonged in the chair and had sat in the spot many times.

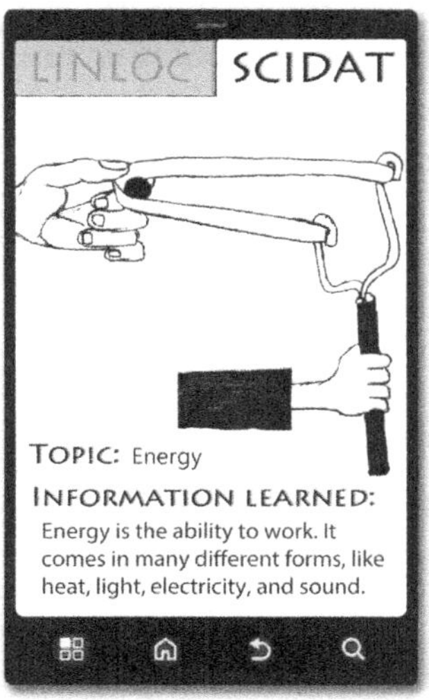

The twins looked at each other and shrugged. Was this their local expert? They thought it might be until they heard more footsteps approaching, and what sounded like someone sharing information about science.

"Energy is the ability to work, and work cannot be done without energy. You can't see energy, but it helps make things happen. It comes in many different forms, such as heat, light, electricity, and sound," the voice said.

The Sassafrases looked out to see four more people approaching. They saw three men—two of which were wearing cowboy hats—and one Brazilian woman.

"Energy frequently changes forms, but it can't be created, and it can't be destroyed," said the man not wearing a cowboy hat. "For example, plants change light from sun into food they can store or use to live and grow. Or electrical energy that flows into a lightbulb and is changed into heat and light. The way energy changes forms is called an 'energy chain.'"

The cowboy-hat-clad men wore gray suits and cowboy boots. The scowls on their faces said they were not enjoying the third man's words on the science of energy. The woman, however, smiled.

"Most of the energy on earth comes from the sun, but we also get energy from fossil fuels, wind, and water." The man's bleached-tipped dark hair gelled into spikes shivered as he spoke. The twins took in his brown leather jacket, jeans, and sparkling white T-shirt that set off his brown skin.

His squeaky-clean white sneakers drew their attention as he continued to speak. "We measure energy in Joules, unless it refers to the energy in food, which is measured in calories. We do this because of James Joule. He was a mathematician and physicist from England. In the 19th century, he showed that the different forms of energy, heat, light, etc., are basically the same."

"All right, McShane, enough of the scientific chitter-chatter," one of the cowboy hat wearing men said with a strong Texas drawl. "Let's just get on the chopper and take our flight up to the falls."

"Aw, c'mon, Colt. You know you secretly love the subject of physics," the man who'd been giving the scientific information said with an undeterred chuckle.

The twins were certain now that he, and not the pilot, was

Augustus McShane, their local expert for this location. The man named Colt, along with the other cowboy, responded to McShane with snorts.

"Okay, all three of you, go ahead and hop on the helicopter and find a seat," the Latina woman said cheerfully. "It's time for your sunset flight over the waterfalls with Ferreira Tours!" She gestured for the three men to climb up into the helicopter, and when they did, they finally noticed the twins already sitting there.

"Well, hello, little doggies," Colt said, taking a seat across from the Sassafrases. "I didn't know there were others signed up for this flight."

"Well, we—" Blaine began but was interrupted by the rest of the adults shuffling around the tight quarters.

Gus McShane sat next to Blaine and Tracey in one of the three seats facing the cockpit. The other two men sat in the two seats facing the twins. The Brazilian woman slid between those two seats through the opening into the cockpit.

"This guy is huge," she said, looking at the pilot and laughing as she took the other seat. "I rarely let my guests bring their own pilots, but he has an impressive resume and the fact that you two are paying more than double what normal guests usually pay—"

"Ole Ox is the best," Colt cut her off. "He's flown choppers all around the world, and we feel most comfortable with him at the controls. Ain't that right, Barlow?"

The other man in boots, evidently named Barlow, nodded. The woman instructed everyone to put on headsets. As they did, the giant pilot started up the rotors. McShane closed the side door as the long blades of the flying machine began to spin.

Blaine and Tracey took off their zip-lining helmets and put on their aviation headsets. Colt and Barlow took off their cowboy hats, replacing them with headsets. McShane put a headset on, mashing down his mullet a bit.

As the skids of the helicopter lifted off the cement circle, Colt was the first to talk into his microphone, enabling everyone to hear him over their headphones.

"Up we go, people. Off to these waterfalls that everyone is talking about like they are so great. At least we are getting to ride a helicopter at sunset, right?" He then looked directly at the Sassafrases. "Now, who are the two of you again?"

"We are Blaine and Tracey," Blaine answered for the two of them. "Blaine and Tracey Sassafras. We are here to learn about physics."

"Of course you are," Colt laughed, but not like he was impressed. "Well then, you two should fit in nicely with this guy," he said, pointing at McShane. "We just met him. Seems like an alright fella, except for the fact that he keeps talking all scientifically, and we're not really sure why."

"I told you, Colt," McShane said with a dauntless, dimpled smile. "I'm a consultant, and I'm here to see the falls and the dam if possible."

Colt smirked at McShane, like he was not impressed or convinced. He looked back at the twins. "Well, I am Colt Belford. And this is my associate, Barlow Hennessy," he said as he reached over and slapped his buddy on the shoulder.

"We are from the great state of Texas. As is Ox, our pilot. We are here to see the falls. And yes, possibly the dam as well."

"And I am your tour guide, Adelina Ferreira with Ferreira Tours," the Brazilian woman said, chiming in from the cockpit. "You two must have booked last-minute online."

The twins nodded and smiled at her as she looked back.

Ferreira continued, "The falls these guys are referring to are the Iguazu Falls. One of the Seven New Natural Wonders of the Modern World. And the dam they are talking about is the Itaipu Dam, which is one of the Seven Manmade Modern Wonders of the World. And yes, I think we will definitely have time to see both Iguazu Falls and the Itaipu Dam. So prepare to be awestruck!"

"And not just by the beauty of the wonders, but let's also allow ourselves to be awestruck by the physics of these wonders," McShane added with a wink. "Hello, Sassafrases, I'm Gus."

Kinetic and Potential

Ox navigated the helicopter over the treetops, hills, and valleys of the Brazilian rainforest. Blaine and Tracey stared out of the large windows, at the slowly darkening green with pops of vibrant color here and there. The ride showcased the verdant vegetation with bold blooms and blossoms.

The helicopter angled back and forth, depending on whether Ox adjusted to the right or to the left as he maneuvered the aircraft. The Sassafrases slid around a little, but they never felt like they were going to fall out of their seats. They were glad the pilot flew close enough to the tropical canopy to afford them such wonderful views.

Suddenly, the green of the forest fell away, surprising the twins. A massive oblong chasm filled with water appeared. Blaine and Tracey literally gasped as they beheld the aquamarine swirls, with its swishing white and turbulent speed. Their ears picked up the thundering of an unfathomable mass of water pouring over multiple steep edges.

"Awestruck?" a delighted Adelina Ferreira asked, looking back at them from the cockpit.

The twelve-year-olds nodded.

"Iguazu Falls," Gus said, looking awestruck as well. "Wow. A true wonder—both visually and scientifically."

Colt Belford and Barlow Hennessy, however, did not seem struck in any way. The two looked like they'd just as soon yawn in boredom than gape in wonder. Something that didn't make sense to the twins because this was truly one of the most beautiful things they had seen.

"Wow," Gus McShane said again. "The beauty. The power. The sheer energy."

"The Tupi-Guarani people who are indigenous to the area named the falls Iguazu," Adelina said in her very professional tour guide voice. "Iguazu means great water. The falls were formed by years and years of erosion from the Iguazu River flowing over basalt rock and sandstone."

Ox flew the helicopter in what felt like slow, wide circles around the breathtaking collaborative waterfalls as Ferreira spoke. "The falls lie partially in Brazil and partially in Argentina. Additionally, the border of Paraguay is not far away. This fabulous place is considered to be a jewel for all three countries. As I mentioned earlier, Iguazu Falls were named one of the natural wonders of the modern world in 2011, and then in 2013 they were designated as a UNESCO World Heritage Site."

"And as you can see," Adelina smiled, "They are especially beautiful at sunset."

"Wow. The sheer energy." Gus repeated before adding, "The potential energy being converted into kinetic energy."

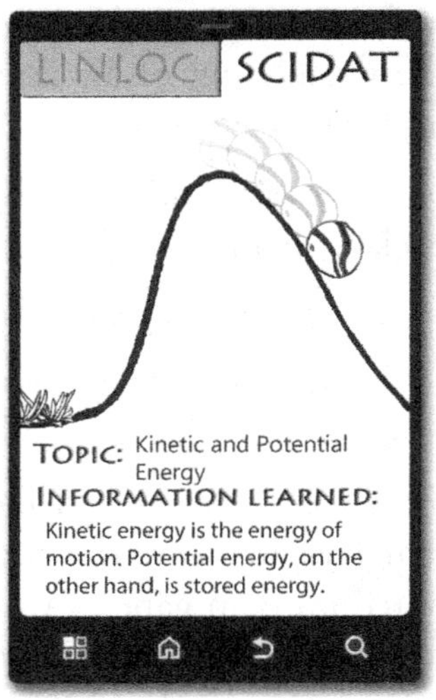

"Kinetic energy?" Blaine asked.

"Potential energy?" Tracey followed.

"Yes," the local expert confirmed. "Kinetic energy is the energy of motion. It's energy that's spent when something moves. For instance, a bird is using kinetic energy when flying. A flag is using kinetic energy when it flutters in the wind. Children running are using kinetic energy. And a roller coaster is using kinetic energy when going down a hill."

Blaine and Tracey snuck a look at each other and smiled at the mention of a roller coaster. They had just experienced that kind of thrilling kinetic energy in Papua New Guinea at the Centripetal Village amusement park. They were also guessing they had experienced kinetic energy at Camp Zip Fire while riding the zip lines.

"Potential energy, on the other hand, is stored energy," Gus continued. "It is energy that is ready to use. Once it is used, it changes into another form of kinetic energy. Some good examples of potential energy are a drawn bow, a coiled spring, a stretched rubber band, a lifted hammer, or a ball at the top of a hill. Even things like the charge in a battery or the stored fat in our bodies are forms of potential energy.

"And this," their local expert said, pointing excitedly out of the chopper's window at Iguazu Falls. "This is showcasing both potential and kinetic energy!"

"How so?" Adelina asked from the cockpit.

"The water at the top of the falls has some kinetic energy and a fair amount of potential energy because of its gravitational position. As it cascades over the edge and thunders down, the potential energy is transformed into kinetic energy as the water moves and gains speed."

The tour guide and the two twins nodded in understanding. The two cowboys did not. Probably because they did not care. Barlow proved this sentiment true when he opened his mouth.

"Who really cares?" The suit-wearing cowboy said with a huff. "Besides all the science stuff, are these waterfalls really all that spectacular? I mean, I guess they're all right, but should they really be one of the seven wonders?"

"Of course they should!" Adelina shot back with a smile.

"Top one hundred, maybe," Colt said with the same attitude

as his associate. "Possibly even in the top seventy-five, but surely you four really don't think the Iguazu Falls should be one of the top seven natural wonders of the world."

"I certainly do," Ferreira stated her same feelings about the matter in a slightly different way. "But I'm Brazilian, so maybe I'm biased. What do you think, McShane? What about you, Blaine and Tracey?"

All three nodded in agreement.

Ox guided the helicopter in a few more wide circles around Iguazu Falls. Then, as the sun fell completely behind the horizon, he broke out of his circular pattern and began flying the aircraft straight in a particular direction. Adelina perked up like she knew where they were going.

"Oh, good! So we have time to go to the dam?" The Brazilian seemed to be asking this question of Ox, but it was Colt who answered. "We most certainly do."

Ferreira turned and excitedly addressed those in the plush leather seats. "I'm so glad we are going to be able to see Itaipu Dam. It is also a wonder to behold! It is a little over fifty kilometers away, and it will be almost dark when we get there, but that is okay because it is so beautifully lit up at night. So much so that many think it is even more gorgeous in the nighttime than it is in the daytime."

Colt and Barlow both perked up a bit. They seemed much more excited to see the dam than they had been to see the waterfalls.

"Just like Iguazu Falls, the Itaipu Dam was also named by the indigenous Tupi-Guarani people. In their language, the word Itaipu means roaring stone," the tour guide explained. "Construction on the dam began in 1975, and the project was completed in 1982. It was a joint venture between Brazil and Paraguay. It is one of the world's largest hydroelectric dams. It curves along almost five miles of the Alto Parana River, and at its highest point, it stands at over six hundred and forty feet tall. Itaipu generates 15 percent of Brazil's electricity and a whopping 90 percent of Paraguay's!"

"Wow! The power. The sheer energy." This time it was not Gus who uttered these words, but Barlow. Something that struck the twins as strange.

"In 1994, the American Society of Civil Engineers named Itaipu Dam one of the artificial wonders of the world."

Adelina stated this fact which was followed up with a hearty, "Yes, they sure did!" by Colt.

McShane, who also seemed excited about seeing Itaipu, shared science. "Dams can create electricity through hydropower. It is a process that starts with potential energy. Water is held in a reservoir. The water is then released through the dam, creating kinetic energy."

"Oh, c'mon man!" Colt complained.

Gus continued, undeterred. "The falling water then spins the blades of huge turbines. The spinning turbines activate a generator. The generator produces electricity. The electricity is then fed into the electrical grid to power homes, businesses, cities, etc. Hydropower is considered to be a renewable source of energy because the energy is coming from a natural source that can be replenished faster than it is consumed. It is also considered to be a source of clean ene—"

"Enough!" Colt actually shouted through the helicopter headsets, interrupting Gus McShane. "That is all the science-speak we want to hear from you!"

McShane didn't seem shocked by this outburst, but the Sassafrases and Adelina Ferreira were.

"Enough of the science and enough of the pretending like we don't know who you are," Colt continued, still mad, but no longer shouting. "We know your full name is Augustus McShane. We also know that you're not just here in Brazil because you're interested in the beauty or science of Iguazu and Itaipu."

McShane remained quiet, not responding in any way other than a slight, knowing smile.

"And you aren't just a mere consultant," Colt kept going. "You work for the Seven Wonder Foundation. They've hired you to travel around the globe doing all their dirty work, fixing all their little problems as they arise."

Other than the sound of spinning blades, there was silence in the helicopter. McShane's knowing smile remained. Colt and Barlow scowled. Ox flew steadily on. Blaine, Tracey, and Adelina were bewildered.

"I also know who you are and why you are truly here," Gus said, breaking his silence. "Your actual names really are Colt Belford and Barlow Hennessy, and that really is your pilot named Ox. You are from Texas, and you came here not to see Iguazu Falls, but to get to Itaipu Dam. Am I right so far?"

Now both Belford's and Hennessey's expressions went from angry scowls to knowing sneers.

"Way back in 2011, the two of you nominated the Cowboy Boot Bluff to become one of the new Seven Natural Wonders of the World," McShane continued, his statement receiving confirmation from both suits.

"Cowboy Boot Bluff is a marvelous natural wonder and one of the many glories of Texas," Barlow stated proudly.

McShane spoke on. "When all the nominations for new natural wonders went through the global voting process that year, Iguazu Falls received overwhelming support. Cowboy Boot Bluff, however, received a total of only three votes."

"It was a tragedy," Colt blurted. "How is it even possible the votes played out like that? Iguazu is just another group of boring waterfalls. Cowboy Boot Bluff is a one-of-a-kind rock formation that looks exactly like a cowboy boot! There's nowhere else in the world like it!"

"That very well may be true," McShane said in a way that gave more than a slight hint of his disagreement. "But the global voters had spoken. In 2011, Iguazu Falls officially joined the list of natural wonders and your beloved Cowboy Boot Bluff did not."

At this, Colt and Barlow both grew visibly angry again.

"All these years the bitterness, resentment, and jealousy have been boiling," Gus said. "And now the two of you—three with Ox—have finally decided to do something about it. You've hatched a plan to exact revenge. Am I right?"

Neither cowboy confirmed nor denied.

"You have hired a convoy of trucks filled with detergent pods to meet you on top of Itaipu Dam this very night. Your plan is to dump all these pods into the water at the dam which will then go downstream to Iguazu Falls, effectively tarnishing these two wonders and turning them into the world's largest bubble bath!?"

The two Texans now sneered.

"That's right, McShane," Colt said. "Those detergent pods will quickly dissolve in all that churning water, causing bubbles and foam to grow and expand at an uncontrollable rate. The bubbles will cover the dam; they will flood downstream and get churned up all over again by the Falls. Yes, it will be the world's largest bubble bath!" Colt and Barlow were both now laughing.

"Let's see what all those global voters will think about that!" Hennessy chirped. "Let's see if they think a giant bubble bath is more wonderful than a bluff that looks like a boot!"

Blaine and Tracey gulped. Adelina still looked shocked. All this animosity had blindsided the three of them. Instead of making awkward eye contact with anyone in the chopper, the Sassafrases looked out of the window. As they did, they saw they were approaching a beautiful, curved line of lights. It had to be Itaipu Dam, and it stood out against the now dark rainforest like a ribbon of shining gold.

Ox flew directly toward that ribbon, and when he reached it, he set the aircraft gently down directly on top of the dam. The tension in the helicopter reached its apex.

Colt Belford looked directly at Gus McShane. "I suppose the Seven Wonders Foundation sent you to try and stop us?"

McShane nodded, his dimple-flanked smile present.

"And how do you think you're going to do that?" The cowboy asked the consultant.

"I don't know," McShane answered. "I usually just figure it out as I go along."

Belford and Hennessy both chuckled arrogantly and glanced up into the cockpit toward Ox.

"What?" McShane shrugged as he reached over and opened the side door of the helicopter. "You think one big guy can stop me?"

Still cackling, Colt now pointed out of the open chopper door. "Maybe not just one. But I bet five big guys can."

Gus, along with all the others, looked out onto the dam where Belford was pointing as five mercenary-looking giants emerged out of the shadows. Their sneers were bigger than Colt's and Barlow's, and they looked more than ready and capable of handling Gus McShane.

CHAPTER 7: THE BELLY OF THE CONCRETE SNAKE

Hiding Heat

The five mercenaries had handled Gus McShane. They had also pulled Adelina, Blaine, and Tracey out of the helicopter. The four had their hands secured behind their backs with plastic zip ties. They walked to the edge of the dam, where the stomachs of the Sassafrases dropped. They thought they were about to get shoved over the edge down into the dark, thundering water below, but thank goodness, that was not the case.

There was a set of solid-looking stairs built into the structure of the dam, leading down to a small arched opening. The quartet was led down a long, narrow passageway. Under normal circumstances, Blaine and Tracey would've loved getting to take a tour of the inside of a world wonder, but right now, this felt more like the unwanted tour inside the belly of a long fortified concrete snake. The corridor was bare except for the length of conduit running overhead and a door here and there. The tunnel smelled musty. All the footsteps echoed loudly off the polished cement floor.

Eventually, they reached a locked metal door. A mercenary pulled out a set of keys, opened it up, and led the bound four inside. The twelve-year-olds looked around. There were a few things in this room, but not many. It looked like maybe it was some kind of old kitchen or dining room with a dilapidated wooden cabinet, an old rusty fridge, and a small crusty electric stove with several stains from spilled liquids that had never been cleaned up.

Colt addressed the four with a relaxed face. "We are not trying to hurt you guys. We really didn't even want to tie you up. But we know that if we don't," he said, pointing at Gus, "He will try to stop us from dumping the detergent into the water."

"I sure would," McShane responded, still seemingly undaunted.

"That is exactly why we are locking you up in the engineers' kitchen," Colt explained.

"The convoy of trucks should arrive in a couple hours," Barlow said. "You are not our prisoners. Once we have dumped the detergent pods and the bubble onslaught has started, we will let the four of you out. Who knows, you might even enjoy watching the bubbles grow and flow downstream toward Iguazu Falls."

With that, Belford, Hennessy, Ox, and the group of intimidating men exited the room and locked the door securely behind them. Their muffled footsteps could be heard walking away from the door back down the long corridor.

"Not their prisoners?" Adelina decried. "Yeah right. They locked us in this place with our hands tied behind our backs!"

The anxiety on the Sassafrases' faces showed they were feeling the same.

Ferreira continued, "And even worse than our situation in here is the situation for my country, and for the countries of Paraguay and Argentina! Those backward gauchos are going to mock and destroy two of our greatest landmarks, not to mention contaminate our water. Gus, we've got to get out of here! We've got to stop them!"

Blaine and Tracey looked at Gus McShane and saw that his dimpled smile was gone, however his confidence was not. There was a look of resolve on his face. They didn't see how all the resolve in the world could get their hands free, bust open that door, and get them out of this place.

Now he just needed to make sure no one could get out of this place. He reached up and wiped his hairless, sweaty brow. He had finished the hard work of transporting the Forget-O-Nator. Next step was to camouflage it, then figure out a way to both lock it and activate it from the inside. More sweat drained from his forehead right into his eyes, making him squint and blink. With no eyebrows to stop the perspiration, the annoying, stinging sweat would often stream right into his eyeballs.

It was dark outside. Thaddeus Wazeezy was standing inside the mobile-home-sized Forget-O-Nator, finally on the grounds of the junior high school. He had been trying for a couple of days to transport the big rectangular machine to the school. That had proven to be a tough thing to do during the daytime. There weren't really many individuals out and about, except that old science teacher of his, Mr. Womberfraggle. He had been at the junior high virtually every day, all day, walking around the building, whistling, smiling, and talking to himself. What was he doing? Didn't he know it was summer? Didn't he know there were no students here to teach? Why was he being so vigilant?

If only the old man would've been this vigilant when Thaddeus was in junior high. Then maybe he'd still have his eyebrows. Maybe he could've lived a normal life instead of a life filled with fury and vengeance. It made Wazeezy want to scream just thinking about it. Yet he held his outburst in and focused on the task at hand.

The Forget-O-Nator was pushed snugly against an outside wall of the junior high. Its door was lined up directly with an exterior door of the school. From the inside of the school, it no longer looked like you were walking through a doorway outside, but rather through a doorway into a classroom. However, from the outside, you could still see the Forget-O-Nator was not a part of the building. So with the hours of darkness still left, Thaddeus would stack and lay brick around the memory wiping machine to match, making it look like a natural extension of the junior high building. When that was finished, he would figure out a way to lock the door and turn on the machine from the inside since it was currently designed to accomplish these functions only from the outside. Once he did, his memory could be erased along with theirs.

He made one more attempt at wiping all the sweat from his brow, and then Thaddeus exited the Forget-O-Nator, making his way outside in order to start his devious masonry project. He just hoped he had enough time to finish all that needed to be done before the time of the 'reunion' arrived.

"Barlow said the convoy of trucks is supposed to arrive here in a couple of hours, right?" McShane said to his fellow captives, all three of whom were probably still a little in shock. "So that's the time we have to figure out a way to get out of here."

The man with the mullet began looking around the room, studying it for the potential usefulness of everything he saw. The

Sassafrases saw an ingenuity in his eyes that suddenly gave them hope. Adelina Ferreira must have sensed these things too because she visibly calmed.

"Blaine, is there anything in the cabinet that can help us get these zip ties off our wrists?" Gus asked the boy. "Like a knife, a can opener, or anything?"

Blaine, who was standing closest to the cabinet, stood on one foot and used his other foot to open both a top drawer and a bottom drawer.

"No utensils of any kind, Gus," Blaine reported. "Just a pot, a pan, some salt. I think those are coffee filters, and then there are a few dozen sugar packets."

"What about the fridge, Tracey? Is there anything in it? Is it even running?"

Tracey, who was closest to the appliance, opened both the refrigerator section and the freezer section, much in the same manner her brother had opened the doors of the cabinet. The refrigerator was on, but it was virtually bare, and it smelled terrible. There was just a bottle of water and a couple of empty glass jars in the fridge and nothing at all in the freezer. She reported what she found.

"What about that over there in the corner? Is that a first aid kit?" Gus asked.

Adelina stepped over to inspect. "It is," she confirmed excitedly. "Maybe there are some of those small bandage scissors in it! That would work to cut off our zip ties.

Ferreira sat down in a position that put the kit directly behind her. She used her bound hands to open up the small plastic box and then quickly twisted around to see what was inside.

"Nothing," she said, frowning. "It's just a handful of Band-Aids and an instant cold pack. No scissors, no Q-Tips, no alcohol wipes, no tweezers."

Even with the meagerness of what they'd found, Gus McShane did not seem discouraged. "Well, the lights are on in the room and the fridge is running, so that's good," he said. "That means we have electricity. Which shouldn't be all that surprising, considering we are inside a facility that was built to generate electricity. But what I am hoping this means is that the electric stove is on as well."

"How is an electric stove going to help us get these zip ties off our wrists?" Blaine asked.

"Heat," the local expert responded as he walked over to the stove and began fidgeting with the knobs. There was a defined click sound. Almost immediately, one of the black coils on the stovetop turned red.

"Yes!" Gus exclaimed. "This will work great! We just need to be careful not to burn ourselves."

"So instead of cutting the zip ties off, we're going to burn them off?" Blaine wondered.

"We're going to try," the consultant confirmed. "It looks like this coil is going to be more than hot enough to do the trick. But of course, the problem is, our hands are all tied behind our backs, so we won't be able to see the stove in order to line up the burning coil with the correct spot on our zip ties. Blaine, I'm going to need your eyes."

The boy gulped as he walked over close to his local expert.

"Set at medium heat, a normal electric stove should heat to a temperature of somewhere between three hundred and fifty to four hundred degrees Fahrenheit," McShane explained. "Temperature is a measure of how hot something is or how much heat it has, and it can be measured in both Fahrenheit and Celsius. The nylon plastic zip ties around our wrists usually have a working temperature range of negative forty to one hundred and eighty-five degrees Fahrenheit. So a coil burning at three hundred and fifty degrees should melt the ties and allow them to snap apart rather quickly."

Blaine gulped again. Tracey also gulped, even though she wasn't the one guiding their local expert's hands toward a burning electric coil.

"Heat is a type of energy," Gus said.

"One inch to your right," Blaine directed.

"Heat energy moves from hotter to cooler places. Something warm gets cold because the heat energy moves from it into the colder air surrounding it."

"Stop. Now down two inches." Blaine guided.

"Conduction, convection, or radiation can transfer heat. Conduction carries heat through an object until the temperature evens out. Such is the case with this electric stove. Convection carries heat through liquids and gases. Radiation carries heat in straight lines through space."

"Okay, now just a half inch back to your left." Blaine moved to the right himself.

"Typically, when a material heats up, it gains energy. The molecules speed up and spread out. This causes the material to expand or get bigger."

"Right there. Don't move." Blaine said.

"On the other hand, when materials cool down, they lose heat energy. The molecules will typically slow down and get closer together. This causes the material to contract or get smaller."

Snap! The zip tie popped. Gus raised his hands over his head, no longer bound.

Adelina cheered. Tracey applauded. Blaine let out an enormous sigh of relief. He had burned none of his local expert's fingers off or set his brown leather jacket on fire.

"Okay, Blaine, you're next. Time to burn your zip ties off," McShane said. "And this time around, it should work even better because instead of just verbally guiding your hands, I can physically guide them to the coil."

Blaine went from sighing to taking another gulp.

The high beams sliced through the Brazilian night, but they didn't provide as much light as he needed amidst all this imposing jungle darkness. With squinting determination, he gripped the wheel firmly at ten and two, vigilantly keeping the huge military-style truck on the narrow, winding, earthen, menacing, mountainous rainforest road. The resolve he was driving with was not only going to serve him but also the convoy of covered trucks that were following him.

There were twenty trucks carrying the ever-important cargo. He was not only leading them because of his huge muscular stature that put him at nearly seven feet tall that seemed to intimidate any potential foe, though that was part of it. The main reason he was at the helm of this nighttime convoy was because of his skill as a driver of large vehicles and machinery.

That coupled with the fact that he spoke a little Portuguese. He quickly glanced at his wristwatch. He had less than two hours to get to the dam. He stepped down a little heavier on the accelerator. They were going to arrive at Itaipu on time.

He would make certain of that.

They had pried it, pulled it, rammed it, and had even kicked it with all of their might, but the substantially solid metal door had not budged at all. Thanks to heat, the four captives were all now free of their bonds, but not of their prison cell. With the ability to now freely use their hands, they had checked every inch of the room, including their pockets, for something they might use to pick the lock on the door, but they had come up empty-handed.

Blaine, Tracey, and Adelina sat on the floor. McShane slowly walked around the room as if he was still assessing every object and its potential usefulness.

"If the timeframe those Texans mentioned is true, then those trucks full of detergent pods are going to be here in a little over an hour," the Brazilian tour guide said, full of worry. "We've just got to get out of here! We have to get out so we can try to stop them!"

"Chemistry!" McShane suddenly exclaimed.

"Huh?" Ferreira quipped.

"Well, chemistry and physics together. That's how we'll do it," the spiky-haired man continued optimistically. "We'll mix a solution and change its state, employing these two scientific subjects. It's our best chance to get out of here, and I think we have just enough time to do it."

"Do what?" Adelina asked.

"Make an explosive powder," Gus answered.

"To blow the door open?" Blaine questioned.

"With what?" This time it was Tracey who asked the question.

"With the stuff we found in the cabinet, the fridge, and the first aid kit," McShane answered.

"Really?" all three simultaneously said.

"Really," McShane said, jumping into action.

He grabbed the bottle of water and the glass jars out of the refrigerator. He grabbed a pot, a coffee filter, the salt, and all the sugar packets out of the cabinet. Gus went over, picked up the first aid kit, opened it up, and took out the instant cold pack, looking it over.

"Good," he said, reading the small print on the packaging. "This has what we need."

He added the cold pack to the other items. "If everything goes well, we can use all this stuff to make a flammable powder that we can put around the lock on the door. Then, we can light it on fire to blow apart the lock and open the door."

Gus focused on what was before them. A confident look spread over his face. The other three faces did not share the same look.

"What is taking them so long to get here?" Colt complained. "I hoped they'd have arrived by now! We could be dumping all that detergent over the side of this dam."

"I know. It seems like it's taking forever," Barlow agreed.

The two wonder wannabes, one large pilot, and five hired mercenaries stood atop the Itaipu Dam under the brightly glowing electric lights, waiting for their shipment of soap.

"But they aren't scheduled to be here for another hour," Barlow reminded. "They'll get here, Colt. They will. That guy leading the convoy had a certainty about him. All those trucks will get here, and we will dump all those pods."

"I'm sure you're right," Colt said, nodding. "And in the morning everyone will wake up to a bubbled over dam and a bubbled over Iguazu Falls!"

"The world's largest bubble bath!" Barlow added.

Colt was now cackling. "Take that and name it one of the Seven Wonders, all you no-good global voters!"

Changes in State

"I'm not going to bring it to a boil," Gus explained as he stood over the electric stove tending to the pot. "I am just heating it up enough to ensure everything dissolves. The solution needs to be as clear as possible."

The Sassafras twins nodded, even though they still weren't exactly sure what their local expert was doing. Gus stirred up the mixture for another minute or two. Then, feeling satisfied it was ready, he carefully poured the liquid from the pot to a waiting glass jar on the countertop.

"Okay, Tracey, go ahead and open the freezer if you would, please," McShane asked of the Sassafras girl, which she quickly did.

Putting his jacket sleeves over his hands so as not to burn his skin on the glass jar, Gus eased the solution from the countertop to the freezer.

"This is where we move from chemistry to physics," the local expert explained as he shut the freezer door. "We have measured out, combined, and dissolved different elements. Now, in order to move to the next stage of making our flammable powder, we need to change the state of this solution."

Blaine and Tracey had taken precise mental notes on the previous three topics so that they could enter all the SCIDAT into their phones later. They listened so they could do the same again as Gus shared about their fourth topic in Brazil.

"Materials can change between solid, liquid, and gas. We call these changes, changes in state." McShane said as he stepped back over to the countertop and started opening sugar packets and pouring them into the second glass jar. "Changes in state include freezing, melting, boiling, and condensing. These changes happen at different temperatures unique to different materials. Freezing is when liquid gets very cold, so cold that it turns from a liquid to a solid. Melting is when a solid gets very warm, so hot that it turns from a solid to a liquid. Boiling is when a liquid gets very warm, so hot that it turns from liquid to gas. Condensing is when a gas gets very cold, so cool that it turns from a gas into a liquid. And just FYI, occasionally, a solid heats up so fast that it changes directly into a gas, which is called sublimation," McShane explained as he poured the contents of the last packet into the jar.

"So, by putting this solution in the freezer, we are helping it to change in state from a liquid to a solid?" Tracey asked to confirm.

"Exactly," Gus answered.

"And why do we need it to be solid?" Blaine asked.

"We need at least the bottom and edges of this solution to crystalize, or become, a solid. We will need those crystals in order to make our explosive powder."

"Will it freeze fast enough?" Adelina asked. "We only have an hour left before those trucks are supposed to get here."

"I saw there was a thermometer in the freezer when Tracey opened it for me," McShane said, still not an ounce of worry in his voice. "A thermometer measures the temperature of either air or liquid, but anyway, the thermometer said the freezer was at zero degrees Fahrenheit. So, for the amount of liquid we have in that jar, I think it may start crystallizing in as little as thirty to forty-five minutes."

The biggest capybara he'd ever seen had just tried to put an end to the convoy. Luckily, he maneuvered the big truck around the large South American rodent without hitting the animal and without going off the side of the road. He looked back in his side mirror. All the following trucks were still with him. None of them had hit the animal or swerved off the road, either. That was a good thing. Another good thing is that they were finally beginning to emerge from the densest part of the forest. Soon, the road would both widen and straighten out a bit. He glanced at his watch again. His convoy was making great time. By his estimation, they would reach the dam in approximately thirty to forty-five minutes.

The four of them were all just kind of staring at the freezer door. Obviously, they knew this wouldn't help the liquid solidify or crystalize any faster, but somehow it made it seem like it would.

Thaddeus Wazeezy didn't know what else was going on in the world tonight, but here at his old junior high school, the brick laying was going great. So great, in fact, that he had thought to himself a couple times maybe he should've committed his life to something like this: building something with his hands instead of having his mind consumed with vengeance. But no, he couldn't think like that. He needed to finish laying these bricks around the Forget-O-Nator so that he could complete the grander task. His eyebrowless opus was composed of crushing all things Cecil Sassafras.

Summer T. Beach didn't know what else was going on in the world tonight, but here in Mrs. Pascapali's guest bedroom, she was feeling rather lonesome. She was missing her best friend. She had always known it to be true, but she was probably feeling it more acutely now than ever. Her life was so much better with Cecil Sassafras in it.

Colt Belford and Barlow Hennessy didn't know what else was going on in the world tonight, but standing here waiting on top of the Itaipu Dam was wearing their patience thinner than it already was. They had paid a lot of good money to have this plan unfold exactly as it had been drawn up. They had successfully decommissioned that consultant with the Seven Wonder Foundation, but what about his driver they'd hired to get the convoy of trucks here full of the detergent pods? Would he make it on time? They were about to find out because he only had a few minutes left to make his deadline.

Paul Sims didn't know what else was going on in the world tonight, but he knew he needed to up his game and perfect his facade. He had never considered Cecil to be the sharpest tool in the shed, at least socially, but his old junior high classmate was starting to catch on. He could tell Cecil thought he wasn't being authentic. Summer was catching on as well. This wasn't good. He was going to have to step up the charm, and it needed to come across as bona fide.

Paul rolled over on the lumpy bed in Cecil's guest room, certain that he was smelling of bacon and mothballs, just like the house did. He couldn't sleep, but it wasn't because of the smell or the lumpy bed. It was because of his predicament. He'd been part of committing a crime. A big crime. He'd stolen items from his own museum. He knew the Swiss Secret Service was eventually going to find out about his involvement. It was only a matter of time. He only hoped to lighten the blow when he got caught. He planned to let the Triple S think he aligned with Cecil and his crew, especially Summer.

He'd been spending a lot of time with her, but maybe he needed to be more creative. Maybe instead of just hanging out with her at Mrs. Pascapali's house, he should take her out somewhere.

Yes! This was a good idea! He should take Summer Beach on a date. Maybe he could even pretend he had feelings for her. This was a great idea, he was sure of it.

If he was officially dating Summer when Triple S showed up, that would keep him out of trouble, right? He nodded to himself even now, thinking surely so. Now, where in this neighborhood was a suitable location for a date?

Socrates and Aristotle didn't know what else was going on in the world tonight. As Socrates taught, "The only true wisdom is knowing you know nothing." But their hearts broke as they stood in the basement watching him work. He wasn't working with his normal bounce, vigor, or joy. Rather, he seemed to work just to fill the time. As everyone knew, Aristotle pontificated, "Pleasure in the job puts perfection in the work." And there was neither pleasure nor perfection happening down here in the basement.

This place no longer looked like a would-be romantic date location. It was back to its normal, messy, cluttered self. But Cecil was not himself. Nor was President Lincoln. The two could tell the prairie dog was worried about the scientist. They wished there was something they could do to help, but as usual, they would probably just stand here and do nothing.

There it was! The Itaipu Dam. They had made it! And on time! And there they were, standing in the middle of the dam, waiting for him and his cohort. He guided his lead truck from the darkness of the road and onto the electrically lit surface of the dam.

"There he is, Barlow, there he is!" Colt exclaimed gleefully. "And he made it on time! We've got our detergent pods. It's time to make some bubbles!"

Barlow jumped up and down in his boots and laughed with his associate. "Get ready for the world's largest bubble bath!"

Even the usually stoic Ox was now laughing.

Now was the moment of truth. Everything up to this point had gone according to plan. The solution had indeed crystallized in the freezer. It had been crushed up and then carefully mixed. The powder had been placed strategically around the door lock. A fuse had been made from rolled up coffee filters. The electric stove had been used to light the fuse, which McShane was currently carrying toward the door.

Was the powder they had made really going to explode? Was it really going to be strong enough to blow the door open? They were about to find out.

The flame was successfully transferred. "Everybody get back!" McShane shouted.

In a matter of seconds, there was a bright spark followed by a loud explosion. Smoke, ash, and small pieces of metal shot out in virtually every direction. The door had been blown open! Physics and chemistry had worked! They were free.

Wasting no time, Blaine, Tracey, Adelina, and McShane raced through the opening and out into the narrow corridor. But were they going to make it out in time to stop the impending sinister bubble bath?

CHAPTER 8: GREENLAND'S FIRST OFFICIAL BOY BAND

Light Landing

"You think one big guy can stop us?" Colt cackled with a dash of doubt and a sprinkle of confusion.

"Maybe not just one," Gus McShane replied as he pointed over toward the convoy of trucks. "But I bet one hundred can."

Then, instead of any kind of detergent pods spilling out of the backs of the covered military style trucks, men did. They were native Brazilian men who looked a lot more intimidating than Ox or any of the Texans' hired mercenaries.

Blaine and Tracey were more confused than Colt—had their local expert been expecting all this? Because of Gus McShane's ingenuity, they'd successfully busted out of their would-be prison. They had raced up here to the topside of the dam to find the expected convoy of trucks had just arrived.

Colt and Barlow hadn't been thrilled to see they'd escaped. But, because their plan to create the world's largest bubble bath was firmly in motion, they chose to ignore the escapees. The lead driver stepped out of the front truck and that's when things changed. He was the man they'd hired to lead this convoy, and he was as tall and muscular as Ox. However, he waved at McShane and said, "What's up, Gus?"

"What in the world was going on here?" The Sassafras twins wondered. "Who was with who?"

"Everybody meet my buddy, Mark," McShane said with the dimple-flanked smile back on his face. "Everyone thinks that as a consultant I work alone, but I don't. Mark is my associate. We go way back."

"All the way back to peewee hockey," the big truck driver named Mark confirmed.

"But we hired a mercenary to lead the convoy," Colt said in a voice now void of any form of cackling.

"I relieved him of his duty," Mark stated.

"The trucks were supposed to be filled with detergent pods." Barlow caterwauled.

"They were," Mark confirmed. "But I exchanged those pods with these friends of mine."

He gestured toward the scores of men who'd just climbed out of the back of the covered trucks.

Ox, as well as the five men who had been helping the Texans, shrunk back.

"How did you find out about our plan?" Colt asked, defeated.

"We have our ways," Gus answered. "The Seven Wonder Foundation has been keeping tabs on you guys for a while."

The consultant laughed. "During this operation, I lost communication with Mark, so I wasn't sure if he was going to show up or if the detergent was going to show up. Sure glad it ended up being him. Now the dam and the falls are safe."

"But what about Cowboy Boot Bluff?" Barlow cried. "Will any international visitors ever come to see it? Will it ever have the chance to be one of the Seven New Wonders of the Modern World?"

"Top 100, maybe," Adelina Ferreira chimed in. "Possibly even in the top 75, but definitely not in the top seven."

Barlow said nothing, but Colt responded by changing the subject. "If you already confiscated all the detergent pods, why did you bring the trucks here? You already knew our plan was foiled."

"The pods were confiscated, shall we say, mid-transport,"

Mark answered. "That being the case, my Tupi-Guarani friends and I were already on the way here so we thought we might as well come all the way out to the dam to let you know the people of Brazil, Argentina, and Paraguay are not big fans of your world's largest bubble bath idea."

The suit-wearing cowboys nodded and then hung their heads with at least some level of remorse.

A few hours later, after everything had quieted down and the Itaipu Dam was just a dam again, Blaine and Tracey found themselves back down in the small engineer's kitchen. What was once their makeshift prison cell would now serve as their launching point for the next ride on the invisible zip lines.

They were just now finishing inputting all the information they had learned about energy, kinetic energy, potential energy, heat, and changes in state, along with pictures from the archive app suitable for each topic.

"Okay, that's that," Tracey said as she pushed send. "All the SCIDAT is off to Uncle Cecil's for him to view on the data screen down in his basement."

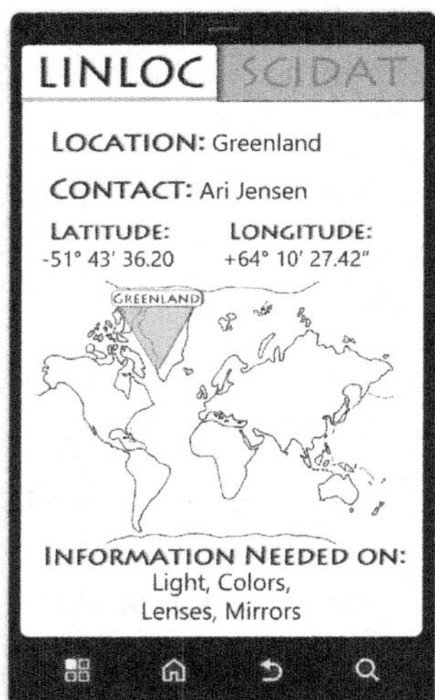

"Now for LINLOC!" Blaine said excitedly.

Where would they be zipping to next? What would be their new topics of study? What adventures awaited?

"Nuuk, Greenland!" Tracey exclaimed. "Longitude 64° 10' 27.42" N, Latitude 51° 43' 36.20" W. We'll be studying the topics of light, colors, lenses, and mirrors.

"With a local expert named Ari Jensen," Blaine added.

The twelve-year-olds put on their zip-line helmets, strapped on their backpacks, cinched up their harnesses, and calibrated their carabiners. Impatiently, they waited the seven seconds between their carabiners clipping into the invisible lines and zip off.

Then zip off it was! Through swirls of effervescent light, from Brazil to Greenland. They landed with the customary jerk and then waited for their sight and strength to return. They were smiling and expectant until their new landing location came into focus.

They had landed directly at the feet of an enormous man. A giant, towering figure who made Grizz, Ox, and Mark look like peewees. Weren't their longitude and latitude coordinates supposed to land them in unnoticed and out of sight places? Yet here they were, landing and now regaining their physical faculties right in front of someone who was huge and holding a large staff. And he was standing over them unmoving and staring down at them . . . kind of like . . . a statue.

Oh, wait. They had landed at the foot of a statue, not a giant person. A person named . . .

"Hans Egede," Tracey read the plaque at the base of the statue. "The founder of the town of Nuuk."

Blaine acknowledged the fact his sister had shared and stood to his feet to take a wider look around. This place was beautiful! The dawn graced the landscape and sky with beautiful hues of red, orange, yellow, and blue. The snowcapped peaks shone in the not-too-far distance. There were pristine bodies of water all around them, even some with small lazily floating icebergs. There were different sized buildings spotted throughout the rocky and hilly topography. The owners had painted most of them in bright and vibrant colors, as if to mimic the sunrise.

There were also . . . people—quite a lot of people . . . heading their way. The Hans Egede statue was on top of the highest hill in the area, and the small crowd headed toward this apex and the twins. A bunch of teenage boys being flanked by attendants and camera crews approached.

Blaine and Tracey didn't know why all these people came their way, but they suddenly felt a need to be ready for their arrival. They quickly took off their harnesses and helmets and put them in their backpacks. Tracey fluffed her hair and checked her breath. Blaine flexed the muscles he wished he had and stood in a posture that only he thought looked cool.

The first two people to approach the top were talking about science, in English, but with Dutch-sounding accents. "A type of energy? Really?" The teenager with stellar blonde hair was asking.

The middle-aged Nordic-looking man, who wore a substantial backpack, hoisted a large camera over his shoulder with one hand. He was carrying an even larger gear box in the other hand and answered, "That's right, Justus, light is a type of energy. It is made up of electromagnetic waves—waves that transfer energy from a source to its surroundings."

"That's crazy," the young man named Justus said as the two climbed closer to the top. "It doesn't seem like that's how it works at all."

"I know," the camera-gear carrying man smiled. "Objects allow differing amounts of light to pass through them," the man, who the Sassafrases strongly assumed was Ari Jensen, continued. "An object or substance that allows light to fully pass through it is known as transparent. An object or substance that allows some light to pass through is known as translucent. When light does not pass through an object, it is known as opaque. We see objects when light bounces off them. The part of light we see is called visible light."

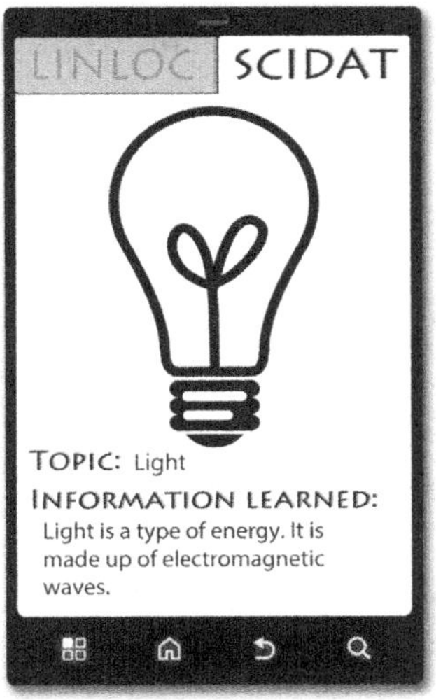

Even with wind swirling around in the area, the teenager's hair remained shellacked into place.

"Opaque objects will create a shadow on the side opposite the light source," the presumed local expert continued. "The umbra is the darkest part of the shadow where no light passes through. The penumbra is the gray part of the shadow around the umbra, where some light reaches the area behind the object."

"Okay, this is helping me understand so much more about what you do, how important it is, and even what some of your equipment does, like those reflectors you were using earlier," Justus commented. Then he asked, "But what does luminous mean? You mentioned that word earlier."

"Objects that give off light are said to be luminous," the science-sharing man explained. "The sun is luminous and is a natural source of light. Whereas things like candles and many of the light bulbs I use are luminous but with artificial sources of light. The brightness of the light is called intensity. The closer you are to the source of light, the more intense the light will be. The farther you are from the source of light, the less intense the light will be. And one more thing you might find interesting: plants can convert light into food and then . . ."

"Hey! Whatever your name is! I told you not to talk to the talent!" A small yet ferocious looking woman shouted as she came up behind the two who'd just reached the hilltop and the base of the statue. Still, no one noticed Blaine or Tracey.

"Rozing, I was just asking Ari about—" the teenager tried to say.

"And enough from you, too!" The intense woman interrupted. "You don't need to be wasting your time talking to any of the support staff. You need to remain focused and remember why we are here. Remember what today is. You're supposed to be the leader, Justus! Start acting the part!"

Tracey was fairly certain both her hair and her breath were presentable. Blaine had to exhale, releasing the flex he had been holding. Yet his posture remained cool, or so he thought. But it wasn't until more teenage boys came up that one finally acknowledged the twins' presence. Well, at least Blaine's.

"Oh, hey little man. You must be the replacement. Right?"

Blaine answered with an "Uh . . ." and then tried to follow up with an "umm" and a "whelp," but before he could get those conversational nuggets out, more people reaching the hilltop interrupted him.

More boys came up with perfect hair, mostly blonde. Most of them were wearing trendy and stylish clothes. There were more camera-men and support staff with them. There wasn't any pushing or shoving, but there was also no camaraderie as everyone crowded around the Stone Egede.

"What in the world is going on here?" The two Sassafras minds were asking. "Why are all these people gathering up here on this hill?"

Maybe their question would be answered by the last person who'd just emerged at the top. He looked to be the oldest in the crowd by many years. He had weathered skin, squinty eyes, and white, not-so-fancy hair. He'd arrived last, but he'd arrived with an air of authority. He climbed up on the base of the statue and addressed the crowd.

"Good morning. My name is Hans Lyberth. I am the mayor of Nuuk, as most of you already know. You also know why we've gathered here today at our founder's statue."

Blaine and Tracey looked at each other and shrugged. They didn't, but they felt like they were about to find out.

"We are here to crown one of your groups as the first official boy band of Greenland."

The Sassafrases' eyes communicated to each other. That is not what they'd thought he would say.

The mayor's statement set off murmurings amongst the small mass. Before the twins could tell if these murmurings were positive or negative, Hans Lyberth spoke again. "As I call out the names of your bands, I want the band members, the talent managers, and any support staff to stand together. Make sure you stay far enough apart so we can easily tell the groups apart."

There were more murmurings from the small crowd.

"N-Step," Hans said, almost as if taking roll in a classroom. Immediately, a group of five almost plastic looking, sharp-jawed teenage boys huddled up together, joined by a few adults and crew.

"Step Sons," was the next band called. They were a mean-looking band wearing mostly dark clothes and with some piercings on brows, lobes, and nostrils.

"Next is Stepladder," Lyberth announced. Now a group of five guys lined up next to each other, each just a little taller than the next, but none quite as tall as their camera crew and support staff.

"Then we have Wolf-N-Step," the mayor said. Another group of five and their crew bunched together. This group had the standard blonde-haired member, but there were also a couple of ginger-haired boys and two dark brunettes mixed in.

"And finally, there is Stependous." Hans said, looking toward the teenagers who were standing closest to the Sassafrases. "Were you able to find a replacement?" He added a question for this last group.

"Yes sir, we were," The boy who'd earlier asked Blaine if he was the replacement answered.

He then reached out and pulled Blaine into their group. The ferocious lady also directed Tracey and Ari Jansen, the cameraman. The Sassafras twins had to look at each other again.

Tracey's expression was a, "I really don't know what's happening."

Blaine's was more of a, "Huh, my dream to be the heartthrob in a boy band is finally happening. They probably chose me because of my extremely cool posture."

The five different bands and their crews were now clearly parceled. Mayor Lyberth continued his announcement. "N-Step, Step Sons, Step Ladder, Wolf-N-Step, and Stependous. Your bands have all been selected as finalists in this nationwide competition to find Greenland's first official boy band. Five groups have made it to this point, but at the end of today, only one will receive the title. Each of your talent managers has been diligently working with you to perfect your dance moves, your roles, your style, and your lip syncing. Today is the day all that hard work will pay off for one band."

Tracey was feeling pretty confused, wondering what was going on. Blaine was beginning to get excited about potentially having the chance to show off some dance moves he was certain would awe any crowd.

"I know up to this point, you have all been through what probably feels like countless competitions," the leader of the town continued. "And today there is one last competition. A competition that will decide the winner. All five bands are considered by the experts to be even. This competition will set one band above the others."

Everyone on the hilltop waited to hear what the mayor was going to say next. "Today, all five bands are going to shoot music videos. The band with the best video will be crowned champions and will become the pride and joy of Greenland."

The Sassafrases felt like maybe the mayor was being a little sarcastic by the way he delivered the last sentence. However, none of the surrounding people seemed to take it that way. They murmured with restrained enthusiasm.

"Greenland is a beautiful country and Nuuk is a jewel in its crown," Hans Lyberth said with absolutely no trace of sarcasm. "We are going to provide all of you with whatever kind of transportation you need in order to travel around to different majestic spots in Greenland to shoot your videos. Spend the day working on your videos. Allow your crews enough time to edit the videos. Then be back here to Nuuk by 7 p.m. We will not be meeting here at Hans' statue, but at his hotel. There we will have both a screen and a stage set up where we will view the videos you shoot and watch each band perform live."

At this, a silent gasp almost seemed to echo over the hilltop. Mayor Lyberth stood his tallest, spoke his loudest. He said, "Work hard, think smart, be creative, and return to Nuuk and the Hotel Hans Egede with the winning video to claim your position as Greenland's first official boy band!"

Colorful Competition

"Take a few minutes to talk and decide as a band about where you want to shoot your video. Then, go to one of the provided vehicles," Lyberth said as he gestured out beyond the hill.

Everyone looked out to see vans and trucks parked at the bottom of the hill. A few boats had docked nearby in the water. There were even a few helicopters coming in. The excitement in the small crowd jumped up another notch.

"Kasuutta! May you have good health and good luck!" The Greenlandic mayor said as he hopped down off the base of the statue.

"Wow! This is exciting," the boy named Justus said as the members of Stependous, plus Blaine and Tracey, turned in toward each other and began discussing this music video they were about to make.

"We could take a helicopter to the top of the Sermitsiaq Mountain," Justus said, pointing to the tallest snowcapped peak that was clearly visible from where they were standing.

"No, let's go further away," another band member with curly, angel-hair-pasta blonde hair countered, but not competitively. "Let's go all the way to the Greenland Ice Sheet to film our video on all that ice!"

"No, that would be too boring," another one of the boys said. "We should go to Qasigiaunquit Fjord near Disko Bay, where we can see whales and have them in the background of our video!"

"Or Uunartoq where the hot springs are!" said the one who'd pulled Blaine in, and then he added another option. "Or Halsey, where the Viking ruins are!"

"Yeah, yeah, or the . . . bergs . . . of ice that are flo—" Blaine began stumbling through a socially awkward attempt at contributing. Thankfully, the feisty lady interrupted him.

"Stop! Just stop! You five are the talent, not the brains. So stop brainstorming about music video locations. That is for me to decide."

The boys conceded as the woman turned to directly address Blaine. "My name is Rozing Kleist. I am the talent manager for Stependous. Therefore, I am now your talent manager. I hope you will be a worthy replacement for Skid. He had brittle bones, but he was a decent enough singer and dancer."

Though Rozing Kleist was only about the same height as Blaine, her fierceness made Blaine feel like she was towering over him. "What is your name, lad?" The intimidating talent manager asked the Sassafras boy.

"Blaine. Blaine Sassafras. And this is my sister, Tra—" Blaine attempted a full introduction before he was cut off again.

"I don't care about any relatives you have," Kleist barked. "Blaine . . . Blaine . . . your name is Blaine," she repeated as if she was mulling over a potential name change. "Well, considering this is all so last second, I guess that will have to do."

She turned back to the other boys. "Each of you tell Blaine your name and your role in the band, starting with you, Justus."

He immediately obliged. "My name is Justus," he said with every strand of hair still in perfect position. "I am the frontman of Stependous."

Next, the boy who'd recruited Blaine spoke up. "My name is Faro, and I am the heartthrob."

Blaine's shoulders dropped a bit upon hearing this. The position he wanted in the band had already been taken.

"My name is Anse," the next boy, who had a couple lines shaved in one of his eyebrows, said. "I am the bad boy."

Tracey heard all the introductions, too. So far, it didn't necessarily seem to her like all the roles fit the individuals.

"My name is Pilu," the last boy, with the curly pasta hair, said. "I am the shy one."

"Okay, then, last and potentially least, is Blaine here replacing

Skid in the role of little brother." Rozing abruptly finished up with the introductions and then repined. "It's too bad Skid broke his leg in practice. By the looks of it, he was a much better little brother than Blaine will be, but we have to play with the cards that have been dealt. I think maybe we still have a chance to win this thing, so let's get to it!"

The short talent manager took a wide sweeping glance over all the provided modes of transportation, and without asking for any input from the band members and without taking into consideration any of their earlier expressed interests, she decided.

"We are going to get on that boat right there," she said, pointing to the largest floating vessel docked down at the water's edge. "We are going to the closest location we can find. We aren't going to waste our whole day looking for some elusive perfect spot. The competition will be won with your succinct dance moves, your perfect hair and make-up, and your roles in the band being played to perfection. Not by any natural wonders being filmed in the background."

Rozing Kleist now turned to Tracey, acknowledging her existence for the first time. "You," she pointed a finger right in the girl's face. "Help him."

The finger now switched to pointing at Ari Jensen and then to the big box, presumably full of gear. "Carry that."

"Yes, ma—" Tracey tried to respond, but was cut off by the next barrage of bossy instructions. "Everyone. Down the hill. Right now. Go quickly, but don't break any bones. I can't afford to lose another one of you. And for the love of Greenland, don't mess your hair up. Go straight to the boat. Don't talk with anyone from the other bands, and of course, don't talk to any support staff."

The boys of Stependous fell in line and followed their manager's instructions. Blaine gave his sister a rueful look as they began making their way down the hill. He wasn't a big fan of the way Rozing Kleist treated everyone, especially Tracey. He also felt bad that his sister had to carry that big box, while all he had to do was make sure his hair looked good.

Most of the other bands and their crews had already left the hilltop with the Hans Egede statue. There was just one group behind Stependous. As Justus, Faro, Anse, Pilu, Blaine, Rozing, Ari, and Tracey carefully hiked down, this group moved ahead of them, too.

"Good luck, Stependous," their frontman said with thick sarcasm as they skipped down and around them. "Or should I say, break a leg!"

The group of five plus their entire crew, which was much larger than Stependous's, laughed at this obvious reference to Skid.

"Who was that?" Blaine asked Pilu, who was in front of him.

"That was N-Step," he answered. "They have been doing much better than the rest of us in all the competitions up to this point. They are for sure the frontrunners to win. We need to shoot a pretty amazing music video today if we want to have any chance to beat them."

By the time the group of eight got to the boat, all the other boy bands had left in their different vehicles by land or air. Stependous was the only group that traveled by sea.

"All of you head to the foredeck," Kleist huffed as they all climbed aboard. "I will go find the captain to give him instructions."

With that, the short talent manager disappeared around a corner, leaving the band and their support staff to find the foredeck all on their own, for which they all seemed very thankful. Justus led the way. Jensen and Tracey brought up the rear.

Blaine was thankful to see that as soon as Rozing left, Ari took the big box back from Tracey and carried it for her. At the same time, he hoisted the large camera over his other shoulder. The Sassafras boy realized their local expert was currently filming and had probably been filming the whole time, from hilltop to boat.

The group walked alongside the gunwale and found the foredeck in no time. The area wasn't huge, but it was flat-surfaced and wide enough. Plus, it was obstacle free. All at once, Blaine was filled with a surprising surge of adrenaline. He ran to the middle of the deck before Tracey could stop him and busted out his best break dancing moves. The twelve-year-old confidently top-rocked into an original move he liked to call the spaceship shuffle. Tracey had regretfully seen this move before and was of the opinion it should probably instead be referred to as the dying kangaroo. Regardless, as she looked on in embarrassment, she had to admit her brother was moving with gusto.

The four other members of Stependous watched Blaine with wide eyes. Tracey couldn't tell if they were shocked, impressed, or alarmed. Blaine finished, transitioning from his shuffling to a series of spins. As he finally came to a stop, Tracey could tell by the expression on her brother's face that he thought he had just wowed everyone. A way too long, awkward pause ensued. The band members stared with wide eyes. Blaine's confidence suddenly jolted, he smiled as he attempted an explanation for his actions.

"So, fellas . . . ha, ha . . . I just thought I'd . . . you know. Show you some of my . . . better moves . . . So, you know . . . you could see what I have to offer . . . to the band . . . and stuff . . . I've always kind of wanted to be . . . in a band . . . or a part of a professional breakdancing crew . . . or maybe like a sick mixture of those two things . . . or something."

Justus took a step toward Blaine as the boy was attempting to finish his long drawn out thought. "Little brother," the frontman said. "That was . . . awesome!"

The other three Stependous members nodded like they agreed. Tracey's jaw dropped slightly.

Anse now spoke up. "I've always wanted to be a radiologist."

"And me, I've always wanted to be a nanotechnologist," Faro, the heartthrob, said.

"I aspire to be an astrophysicist someday." Pilu, the shy one, announced.

"I love science and would love to pursue a career in nuclear engineering someday," Justus added.

Tracey's jaw dropped further.

"So, if you really want to be a breaker, go for it, Blaine," Justus continued. "That's not really the dancing Roz has us doing, but I can tell you have a lot to bring to this competition and to this video shoot we are doing. Being a part of this band isn't necessarily the first choice for any of the rest of us, but we still want to do our very best today."

"Wait, none of you even want to be in this band?" Tracey asked.

"I'm not saying that," Justus explained. "This is a really great opportunity. For right now. I just think we are all dreaming of something a little different for our futures."

The others nodded again.

"We've been training hard and we are absolutely going to do our best today," Faro said. "Make no mistake, we want to win this competition."

"If nothing else, we want to beat N-Step," Pilu added, which garnered agreed laughter from his bandmates.

"So," Justus said, speaking again. "Let's execute our song, lyrics, and our dance steps perfectly. Let's shoot the coolest music video possible. Let's make Stependous look good. And let's make Greenland look great!"

The frontman then turned to Ari. "And all this should be easy enough because we have the best cameraman in the northern hemisphere! Even cooler than that, he's more scientifically knowledgeable than anybody else we know."

"Well, I don't know about that," Jensen said, laughing. "But

I do love my job, and I love the science behind it. I just wish my crew was a little bigger so we could shoot a more comprehensive video for you guys."

"Why is it just you?" Tracey asked. "Some of the other bands had pretty large camera crews and support staff."

Anse raised a lined eyebrow and then said in a whisper. "It's because of Roz. She scared everybody else off."

Everyone agreed.

"But that's okay," Ari said optimistically. "Because now you're here, Tracey and Blaine."

Tracey had a surprised look on her face. Blaine showcased an expression of dewy confidence.

"Blaine, you complete the band and seem to have some workable skills," Ari Jensen expounded. "And Tracey, you can help me by working the second camera. That way, we can shoot multiple angles simultaneously.

"Where is the other camera?" The Sassafras girl asked.

"In that big box you've been carrying," Jensen smiled. The camera man bent down, opened the big box, and began pulling out a second camera along with other assorted gear.

"What ideas do you have for this video?" Justus asked the man. "Sure, let's include some cool natural spectacle, but how can we also include some science?"

"I was thinking of a video with lots of color," Jensen answered.

"Color?" Faro asked. "How is color scientific?"

"Colors are actually different wavelengths of visible light," the Greenlandic local expert started. "We see colors when those wavelengths bounce off an object and our eyes take the light in for our brains to process. When something looks green, it's because the green wavelength of light bounces off the object while the other wavelengths of visible light are absorbed by the object."

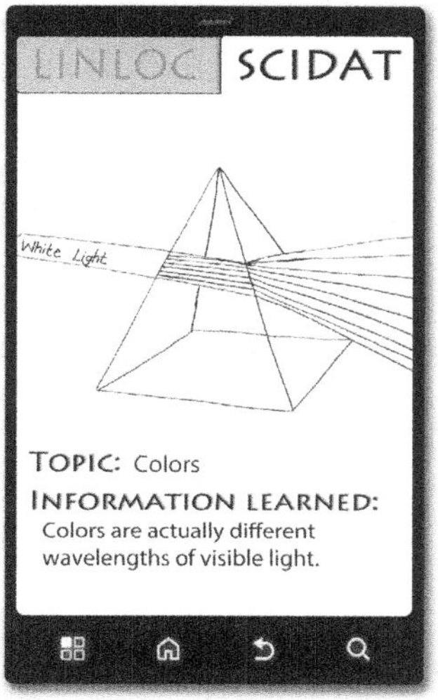

"Really?" Justus asked, sounding just as fascinated as the Sassafras twins were feeling upon hearing this scientific explanation of colors.

Ari Jensen nodded and then continued. "Dispersion is the splitting of light into the visible spectrum. For example, a prism disperses light so that we can see the rainbow of colors in visible light. Visible light appears white, but it is really a combination of seven different colors: red, orange, yellow, green, blue, indigo, and violet. We call these the chromatic colors."

Stependous were all nodding in understanding with Tracey.

"Then we have two ways to describe primary colors: subtractive colors and additive colors," the crew member continued. "Subtractive colors are made by mixing substances. These are the most commonly known primary colors: red, blue, and yellow. When substances are mixed with those colors, they form the secondary colors of orange, green, and purple. Additive colors, on the other hand, are made by mixing light. Almost any color can be made by mixing red, green, and blue wavelengths of light, which is why we call these primary colors of light. Secondary colors of light, like cyan and magenta, are made by mixing the primary colors of light."

"I love the idea of having a cool music video with lots of color. And I think I understand what you are saying about color and light," Blaine interjected, sharing his opinion like he was indeed an established member of Stependous. But he still had questions. "But what about the cameras? I mean, how do they work? How do they, you know, kind of like grab all the colors out of midair?"

Jensen smiled and said, "That's a brilliant question, Blaine."

The local expert started an explanation. "Cameras capture color by splitting light into its red, green, and blue components and then recording the strength of each color on the pixel level. The camera then uses a process called interpolation to calculate the actual color of each pixel by combining the color it captured with the colors captured by the surrounding pixe—"

"What is the meaning of this?" The cameraman was interrupted by a flustered, furious-looking talent manager—a talent manager who was being flanked by an even angrier-looking ship captain.

CHAPTER 9: STEPPING IT UP

Languishing Lenses

Tracey had heard these words so many times by this point, all she could do was roll her eyes. The words were being played over speakers repeatedly as the group tried to sync up both their dance moves and their mouths to the song.

There is a door standing at the top
The top of these stairs so lonely
Would you be willing to take these steps?
If only, if only, if only
Because girl, that door is open
So don't turn around, don't depart
Please step on up and step on in
For that door is to my heart

The name of the song was "Step into My Heart." When Tracey had first heard it, she'd thought it was sweet and heartfelt.

However, as the words and the tune had played over and over again, the Sassafras girl had grown very tired of it. It didn't help when Rozing Kleist, the short ferocious talent manager, stopped the song to shout at the boys about missteps or mistimed words. She also kept going on and on about them not channeling their roles strongly enough.

Ari Jensen had told Tracey that "Step into My Heart" was not original to Stependous. The song had been written by a lyricist for all the boy bands to practice. Meaning that right now, N-Step, Step Sons, Wolf-N-Step, and Step Ladder were also practicing this song and using it in their music videos. The names for each band were also not chosen by the boys but rather by some unnamed executives in some office somewhere. They decided every band should have step in their name so it would match the song. Eventually "Step into My Heart" would be polished, packaged, and sent out by the executives to the waiting masses. The executives were sure Greenland's first official boy band would have appeal far and wide.

All of this served to make Tracey even more weary of hearing it over and over again. Blaine, however, seemed to be loving this whole experience.

Tracey laughed at her brother as he sang along and executed his assigned dance moves as the band's little brother. All of Rozing's admonishment didn't seem to be bothering him at all. Blaine was living the dream.

However, some of the other boys seemed like they were being affected by the talent manager's negative nitpicking. Tracey just hoped none of them would get too angry.

The captain, who was currently steering the boat, was angry enough for all of them. He was so frustrated at Kleist that Tracey thought the old man might actually throw the short feisty woman overboard. He was getting paid for this gig so he would take tyrannical landlubber where she wanted to go, but he wasn't at all happy about it. He'd disappeared back to the bridge as soon as he'd checked out the band and their crew.

There is a door standing at the top
The top of these stairs so lonely

Tracey sighed as Rozing started the song back up again. She would probably end up stopping it soon. They had a deadline for shooting this video, and they hadn't even started yet. Rozing didn't want any filming to happen until the boys of Stependous had every single detail exactly perfect according to her achingly critical eye. By Tracey's summation, that was never going to happen. And so, the boat sailed on to the spot the talent manager had directed, and the practice didn't stop.

The cameraman and his only crew member sat off to the side, waiting for a green light. Ari Jensen had gotten the second camera out of the box and was showing Tracey how to use it. But none of these instructions would matter if Kleist wouldn't have the captain stop this boat to let them get out and film.

Ari gazed up at the clouds and said, "Did you know that diffraction is the bending of a wave when it meets an obstacle or passes through a gap? For example, the way the light is scattered under a cloud at sunset or sunrise. That's from diffraction."

Tracey nodded a no. The minutes and hours were ticking away. Sunset was coming and their 7 p.m. deadline for getting the music video turned in quickly approached. They still hadn't started shooting yet, much less editing.

Ari looked back at her. "Do your best to protect the lens of the camera, and never let it make contact with anything," Jensen said. "Scientifically speaking, lenses are transparent substances with curved surfaces. They allow light to pass through and bend in a particular way."

"I'm pretty excited about getting to use this thing . . . hopefully," Tracey said, trying to think positively. "I really love learning about the science behind stuff."

Tracey's words encouraged Ari, so he shared more information. "Lenses refract light," he said. "Refraction is the bending of light as it passes through different substances. The two types of lenses are convex lenses and concave lenses."

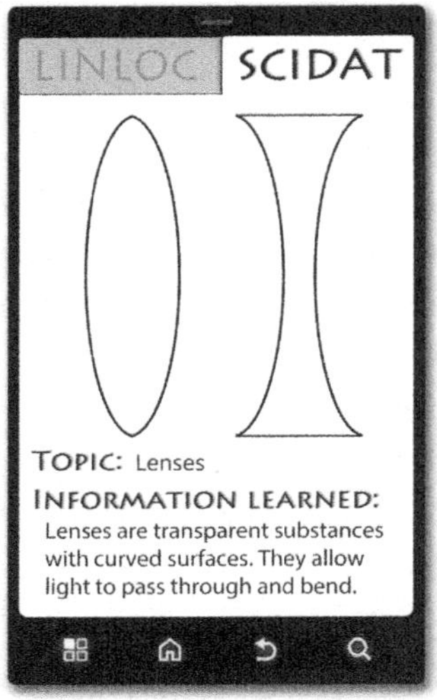

Jensen gestured to the camera in Tracey's hand and continued. "Convex lenses curve outward on one or both sides. These types of lenses are called converging lenses because they cause light to come together at one point. Magnifying glasses, for example, use convex lenses. Concave lenses, on the other hand, curve inward on one or both sides. These types of lenses are called diverging lenses because they cause light to spread out and away. Telescopes are a good example of something that uses concave lenses. Different kinds of cameras can employ both of these kinds of lenses in a variety of ways."

"Hey! Support staff!" Rozing shouted at the two holding cameras as she stopped the song again. "You're talking too loudly. We can all hear you over here, and that's not good. You're supposed to be seen, not heard!"

Ari nodded Rozing's way, acknowledging her, but at the same time not letting what she said negatively affect his attitude. Tracey admired that. Blaine nodded at Tracey as if to say, I just heard that information loud and clear. The other members of the band also seemed unaffected by the boss lady's scathing ways.

There is a door standing at the top
The top of these stairs so lonely

"She started the song again!" Blaine exclaimed internally. At the beginning of the song, Blaine's spot was in the back left. Rozing had them standing in a V formation with Justus at the front. Faro and Blaine were behind him and staggered to his left. Anse and Pilu were behind him and staggered to his right. At first, all the choreographed dance moves looked extremely difficult, but Blaine quickly realized it was just a bunch of stomping, reaching your arms out toward where the camera would be, and contorting your face to make it look like you were happy about having a headache.

Justus was a great frontman. He knew all the moves and the words to the song. Even with all the nagging from Rozing, Justus made Blaine feel like it was going to be possible to nail this thing. The others were all really good, too. The Sassafras boy wasn't really sure why the talent manager kept stopping the song, but no matter, he was having a blast. Even as he thought this, Kleist stopped the song again.

"Concentrate, people! If we want to win this competition, we must be precise! Justus is the only one who sings the first three lines of the song, so no mouth movement from any of the rest of you until line four. Now, let's go again. From the top."

Blaine took a deep breath, squinted his eyes, puckered his lips, and made sure he was in his correct spot.

The music started.

There is a door standing at the top

Arms to the side, slowly lift head, keep eyes squinted, and lips puckered. Blaine repeated the instructions he'd been given for this first line of the song in his mind.

The top of these stairs so lonely

Stomp left and begin lifting arm up, forward, and out to the left.

Would you be willing to take these steps?

Now, stomp right, stomp left, look forward with longing—open eyes a little but keep lips puckered.

If only, if only, if only

Blaine and the others joined Justus, singing along. This fourth line also called for rapid stepping in place and an explosion of both arms reaching forward.

Because girl, that door is open

Now point with right arm. Stop stepping in place. Another swing of the left arm up and out.

So don't turn around, don't depart

Box step into a leftward 360-degree spin, arms in the preparing to spar position.

Please step on up and step on in

High knees, two left and two right, and arms reaching forward again.

For that door is to my heart

Lean forward and look straight ahead. Use left index finger to point directly at heart. Pause and hold. Blaine held his final stance and did his best not to breathe as he squinted and puckered again. Faro, Pilu, and Anse were doing the same. Justus, whose final stance was a little different than everyone else's, was on his knees at the front of the 'V' with both arms raised, looking straight up.

Then something happened that had not happened since they had begun practicing on this foredeck hours earlier. Rozing Kleist remained silent. The talent manager didn't say a word at the conclusion of the song's eighth line. Also surprising to everyone, her ferocious scowl raised one millimeter, maybe even two. Was she about to compliment them?

"That will do," were the words that came out of her mouth, followed by, "Now let's film it."

Ari and Tracey, who were still sitting off to the side with their cameras, were almost too shocked to move. Had she really just said what they thought they'd heard her say? They were actually going to start filming?

"You heard me! Let's go, people. Let's film this thing!" Kleist said, ferocious scowl firmly back in its normal position. "Support staff, where are you? Let's go! Get those cameras over here! Let's get rolling! Justus, Faro, Anse, Pilu, and Blaine, get back into your starting stances!"

Justus asked, "Are we almost to our destination? Shouldn't we start filming when we get to our filming location? Like on a mountaintop or a glacier? Maybe on the shore of a fjord with schools of whales or icebergs floating in the background? Or mayb—"

"No!" Rozing shouted. "We will not be going to any of those places. We will film our video right here on this boat!"

"What?' Pilu puffed, aghast. "If you weren't planning on taking us to any kind of location, why did we even get on the boat?"

"Because I knew the foredeck would be a great place to both practice and film."

"But wha—" the curly haired boy tried to speak again but was promptly interrupted.

"No more from you, Pilu. You aren't even supposed to be talking. You are the shy one, remember?" Pilu now looked even more aghast.

"And no more comments from the rest of you either. You are my band. This is my plan. We are filming right here, right now, on this boat. No ifs, ands, or buts. Justus, Anse, Pilu, Faro, and Blaine, get in your starting positions! Camera crew, get over here and start filming!" The talent manager screamed.

The band members moved reluctantly, but they all made it to their spots. The five boys formed the V at the center of the foredeck. The two-person camera crew moved to the point of the V.

"As I have tried to drill into you from the very beginning," Rozing Kleist thundered, "This competition will be won by precision in dance moves, lip synchronization, and nailing your roles. Nothing else! So put thoughts of whales, icebergs, and mountaintops out of your heads. Instead, hit every move, sing every line, and be the frontman, the bad boy, the shy one, the heartthrob, and the little brother!"

The short ferocious woman then addressed the support staff. "You two, do whatever it is that you do, and do it with perfection!"

Ari Jensen again nodded in his humble yet stout way. He promptly looked into the eyepiece of his camera, and Tracey followed suit.

Everyone seemed to take a deep breath and then the song started up again.

There is a door standing at the top
The top of these stairs so lonely
Would you be willing to take these steps?
If only, if only, if only
Because girl, that door is open
So don't turn around, don't depart
Please step on up and step on in
For that door is to my heart

During the eight lines of the song, Tracey had stood very still at the V's point, holding her camera steady as Ari had directed her to do earlier. Ari himself had started at the same point but had then moved around the group getting mobile shots from different angles. During the eight lines of the song, Rozing Kleist had stood like an angry statue with her arms folded and her scowl in its normal position. But the five boys of Stependous had completely nailed the performance. Every dance step was perfectly timed, every word was perfectly synced, and every role was perfectly played. The music video had been filmed in one take.

"That's a wrap!" Kleist announced with no enthusiasm. "I

will go to the bridge and tell the captain to take us back to Nuuk. In the meantime, stay loose and stretched. Keep yourselves in the right headspace so you can keep playing your respective roles. And for Egede's sake, make sure your hair looks good!"

She then turned toward Ari. "And you! Get that video edited before we get back. Am I making myself clear?"

"As mud," Jensen replied.

The short woman then stormed away from the foredeck to go talk to the captain, who everyone was sure would not enjoy the conversation.

Musical Mirrors

The boat ride back to Nuuk took a couple hours. Rozing Kleist did not end up coming back to the foredeck during the return trip and again, in her absence, there was an almost visible relaxation that spread over the group. They could all rest a little bit and enjoy some of the grandeur of their natural surroundings as they floated by. They could also enjoy a couple free hours of not being shouted at or bossed around.

Everyone except Ari Jensen that is. His time was not freed up. The cameraman had pulled a laptop computer out of his backpack and had spent the entire trip back to Nuuk working hard to edit the music video that would hopefully help Stependous win this competition and become Greenland's first official boy band.

They got back to Nuuk with plenty of time to spare before their 7 p.m. deadline. As they docked, Rozing Kleist reappeared in all of her abrasive glory and quickly escorted her subordinates from boat to land. The surly ship captain also reappeared, just for a moment, with flexing folded arms and a fiery furrowed brow. It was a long enough moment for the Sassafrases to spy a slight smile on his leathery face. It was a smile the twins were sure was present because he no longer had to deal with the Kleist.

The talent manager led her troupe from the water up into

the town of Nuuk toward Hotel Hans Egede, where the video presentations and final performances were slated to happen.

"Hey look, there's Wolf-N-Step in front of us," Pilu whispered, referring to the multi-colored-hair group.

"And there are the Step Sons getting out of a van," Said Anse quietly, pointing toward a parking lot.

Stependous watched the darker broody group alight from their vehicle. That's when they also saw the members of StepLadder pull up and get out of a convoy of trucks they had been using. One got out after another, each a little taller than the one in front of them.

Then, all four of the groups that had already arrived turned toward the sound of approaching helicopter blades. The aircraft drew near, hovered, then landed. The side door opened up and out stepped the plastic-looking members of N-Step with expressions on their faces like they'd already won the competition.

"Well, I guess that means everybody's back and everybody's on time," Faro said, half under his breath.

The five members of Stependous and the two members of the camera crew had stopped walking to watch the landing of the helicopter, but the one talent manager had not. She stopped now and turned toward them, scowling. She opened her mouth for what Blaine and Tracey were sure was going to be quite a shellacking, but instead it was Justus' voice that rang out.

"We're coming, Roz," he said to the manager. Then to his bandmates, "Let's go, fellas. Let's go win this thing."

The group of eight made it to the Hotel Hans Egede where an outdoor stage and screen were both set up.

"This is where the huge crowd will soon begin gathering," Kleist announced. "The masses are waiting for and anticipating the announcement of Greenland's first official boy band. And when that announcement is made, it's going to be Stependous. Why?

Because you were all naturally the best? No. Because I've made you the best. I've taken what little you had to offer, and I've molded it into greatness."

Tracey shook her head upon hearing the talent manager's pompous statement. She sure hoped Rozing hadn't seen her head shake, though.

"This competition will not be won by silliness or scenery," the imperious woman continued. "But by perfect dance moves, perfect lip syncing, and perfect playing of roles. I have given you these things. I have made you perfect because I have forced you to be perfect. So I want you to go out there tonight and be perfection!"

Again, this was a little too much for Tracey, but as she looked at the boys, that's not what their faces were saying. Justus, the frontman, had a face especially full of resolve.

Kleist now pointed toward the Hotel Hans Egede. "They have set up dressing rooms in the hotel. So go now! All of you! Go get ready for the most important event of your lives!"

The members of the band responded with determined grunts and immediately hopped to their leader's instructions. Their support staff followed right behind them. Rozing, however, stayed put at the stage area to chat with some of the other talent managers that had begun to gather.

Once inside the hotel, they were directed by staff to a long hallway that was filled with lightbulbs and mirrors. There were not really dressing rooms as Rozing had said. Instead, it was one long corridor all the bands would share to get ready. The boys of Wolf-N-Step were already here, sitting on stools in front of mirrors, making sure their strands were styled.

Justus found a stool and a mirror before anyone else in Stependous and started looking at this appearance, focusing on his hair, which the Sassafrases didn't think needed to be fussed over because somehow, someway, his hair had remained perfect throughout the entire day. The other three original Stependous members also sat down. The shy one looked good. The bad boy looked even better. The heartthrob was of course looking the best. Yet they all stared at their mirrors and primped.

Blaine sat down at a mirror last. When he looked at the image staring back at him, his first thought was, "Yep, I should've replaced Faro as the heartthrob." Then, he wondered how many fans in the huge crowd tonight would be screaming his name and wanting his autograph.

Ari and Tracey put their camera equipment down and found a couple seats that weren't facing mirrors.

"It's so funny to see so many boys messing with their hair in front of so many mirrors," Tracey laughed.

Ari Jensen agreed with a chuckle of his own. "Would you like to hear about the science behind mirrors?" He asked.

"Sure!" Tracey responded with an echo from Blaine who was sitting close to the two. "A mirror is a shiny surface that reflects nearly all the light that hits it," the local expert started. "Reflection is the bouncing of light off an object. Mirrors are reflectors of light."

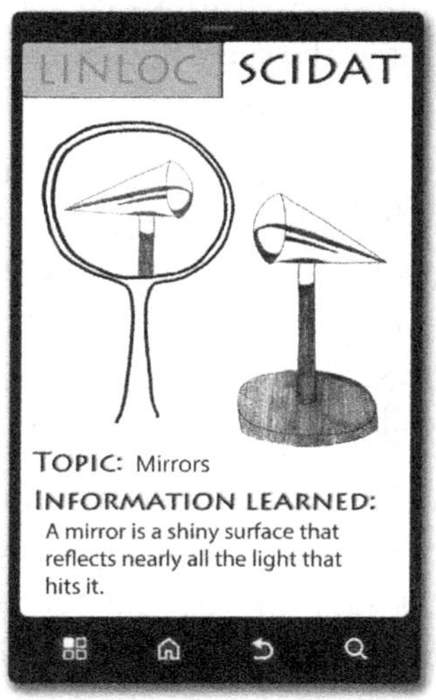

"They are also reflectors of handsome," Blaine said as he worked on making his hair look luscious. Tracey smiled and rolled her eyes.

Ari smiled and continued. "When mirrors are flat, the light is reflected back, which means the image in the mirror is the same size and same up or down direction as the object. The only difference is that the image is reversed, meaning the left and right sides are switched. These are also known as plane mirrors."

"Luckily, my right and left sides are both handsome," Blaine crooned. "So it's all the same over here."

"Oh c'mon bro, for real? You've gotta stop," Tracey said, rolling her eyes again while she smiled and tried not to laugh.

"On the other hand, curved mirrors reflect light at an angle, which distorts the images in them," Jensen kept going with the information. "There are two kinds of curved mirrors: convex and concave. Convex mirrors curve outward, reducing the images' size. These mirrors reflect the parallel lines of light outward. A rearview mirror in a car is a good example of a convex mirror. Concave mirrors curve inward, enlarging the images' size. These mirrors reflect the parallel lines of light inward. Shaving mirrors or make-up mirrors are concave mirrors."

"You mean like this thing right here?" Blaine asked as he pulled at a smaller circular mirror that was situated to the side of the large flat rectangular mirror and was attached to the wall with an extendable metal arm.

"Yes. Exactly," Ari Jensen confirmed.

"Sick," Blaine said in a way that let Tracey know he was trying to act cooler than he actually was. Her brother then proceeded to turn the concave mirror to allow him up close looks up his nose.

Tracey couldn't even roll her eyes anymore. Now all she could do was sigh and shake her head.

As Stependous and Wolf-N-Step continued to prepare their appearances, both Step Sons and Step Ladder came in, found open mirror-facing seats, and began to do the same. As it turned out, the mean-looking boys of Step Sons weren't actually mean at all, and their suspicious piercings were all really just magnetized jewelry. The boys of Step Ladder, however, kept their form by sitting in seats next to each other in order of height. The interactions in the long corridor of mirrors between the competing boy bands were all pretty civil until N-Step arrived. The completely blonde group paraded in with their support crew, which was five times larger than any other group's. They basically took over the mirrored hallway.

"Out of our way, Step-losers," one said to the boys of Step Ladder.

"Don't lose any of your fake jewelry," another said, pushing past the Step Sons.

"Step out of the way, Wolf-N-Step," yet another said to the multi-colored-hair group.

Then, the one who'd laughed at Stependous about breaking a leg earlier that morning, shoved his way forward and found a seat right next to Justus. "What's up, Step-idity? How'd the filming go today?"

The Stependous frontman did not answer. Instead, he just continued to quietly work on his hair.

"What's wrong, man? Cat got your tongue? Not feeling confident, are we? I thought you were prepared to go out there tonight and 'be perfection?'" He said, mocking the words Rozing Kleist had implored her band with earlier. "Ha, ha, no matter. Even if you were feeling confident or perfect, it wouldn't help you. Why? Because your group, along with all the others, are going down tonight. N-Step will become Greenland's first boy band. The rest of you are toast!"

The snooty boy laughed and then turned his attention away from Justus to his own image being reflected back at him by the mirror he was sitting in front of. The artificial-looking kid then snapped his fingers and a host of attendants gathered around him. This kid's cockiness, along with the arrogance of N-Step as a whole, was really frustrating to the Sassafras twins. They guessed most of the others were feeling the same. Regardless, not one member of any of the bands said a word to any of the N-Step boys or their entourage. No scuffles, squalls, or hullabaloos broke out.

Step Ladder, Step Sons, Wolf N Step, and Stependous all finished up, leaving the prepping corridor without incident. N-Step could have the mirrors all to themselves. These four bands would battle N-Step on the stage, with their dance moves and music videos. Not with their words or fists.

A little while later, all five bands found themselves in a curtained off area waiting to go on stage. They were outside, as was the stage and the screen. It felt late to the Sassafrases, but the sun was still up. It must set late here in Greenland in the summer. Mayor Hans stood on the stage and addressed the crowd that had gathered to witness the epic conclusion of this boy band competition.

The bands, crews, and talent managers had been informed of the order they were taking the stage, and Stependous was up first. Then Step Sons, then Wolf-N-Step, next Step Ladder, and finally N-Step. Right now, Rozing Kleist was standing at the gap in the curtains, holding that gap closed, and glaring at her band as if some last second austerity would motivate them toward perfection.

Justus, Anse, Pilu, Faro, and Blaine were ready to go. As soon as the mayor announced them, Rozing opened up the curtain passage, and they would go out on the stage.

They would take their place in the V formation. They would wait for the pre-recorded music to start. Then they would dance. At the same time, their music video would play on the screen behind them. Would it be good enough to become Greenland's first official boy band? They didn't know.

"And now," Mayor Lyberth said into the microphone. "Our first band of the evening. I present to you: Stependous."

CHAPTER 10: THE BACK ALLEYS OF BARCELONA

Wandering Waves

There were only like . . . eleven people in the crowd. The crowd wasn't the large rambunctious mob they'd all been expecting; although, the ones there seemed fairly impressed with the precise steps and movements of Stependous. The quality of Stependous's music video was excellent thanks to Ari Jensen. However, the onlookers seemed pretty bored with the content. The scenery of the video hadn't wowed them.

Justus, Anse, Pilu, Faro, and Blaine knew they had done their best. But none of them were anticipating a win here tonight, and neither were Tracey or Ari. Rozing Kleist, however, still looked certain of victory.

The Step Sons were the second group to take the stage. They shot a video on the Greenland Ice Sheet, which looked cool contrasting with their dark clothing. The next group, Wolf-N-Step, had shot a video next to Qasigiannguit Fjord with icebergs and whales making appearances in the background as the group sang and danced. The small crowd loved that. The next in the sequence, Step Ladder, had shot at some Viking ruins in a place called Hvalsey. They, of course, had arranged themselves in order of height as they danced and sang on centuries-old stone walls.

Then, last but not least, N-Step had taken the stage. The small size of the crowd had not stopped them from performing like they were dancing and singing for thousands. Their competitors had to admit that despite their attitudes and plastic appearances, they were good. Their dance moves were more complex and they nailed every single one. They lip-synched so precisely with the pre-recorded song it looked like they were actually singing. And they weren't just playing their roles in the band—it was like they had each become their characters.

To top it all off, their music video was out of this world. They had done, in reality, what Justus had only been allowed to muse about. They had taken their helicopter across the water over to the base of Mount Sermitsiaq. Then, they sang "Step into My Heart" as they stepped up that snowcapped mountain and into the hearts of the eleven fans who'd gathered.

Mayor Lyberth's announcement confirmed what everybody already knew. N-Step had won the competition. They would be Greenland's first official boy band.

As for everyone else, Mayor Hans Lyberth had seemed relieved the spectacle was over. Tracey had felt the same as the mayor. Step Sons, Step Ladder, and Wolf-N-Step had all hung their heads and walked away like they were disappointed. Ari Jensen seemed to have a mix of those feelings: relieved he didn't have to work for Kleist anymore but disappointed he hadn't been able to help his band win.

The boys of Stependous, however, all seemed relieved. Justus, Anse, Pilu, and Faro had all reacted to the announcement with huge smiles on their faces. Now, instead of being in a boy band, they could pursue futures in science! This encouraged Ari greatly. But then there was Rozing Kleist. The ferocious talent manager had reacted with silence and folded arms. She had simply walked away without saying a single word to her band or the support staff. Only Blaine showed disappointment with the news. His dream of singing and dancing for screaming fans who wanted his autograph would have to be put on the shelf for now.

"C'mon, bro. Quit staring at yourself in the mirror, and let's find some good pictures in the archive app!" Tracey implored her twin brother.

The Sassafrases had slipped away from the others and were now sitting all alone in the corridor of mirrors inside the Hotel

Hans Egede. They had entered all the data they'd learned from Ari Jensen about light, color, lenses, and mirrors into their SCIDAT applications. Now they just needed to add images to that data. Then, they could send all the information electronically to Uncle Cecil, but it was all Tracey could do to get her brother to concentrate on this task instead of admiring his own reflection.

"I could've been great, Trace. I could've toured the world with Stependous, and then maybe someday broken off and started my own solo career. I mean, look at this face. Look at this hair. Lo—"

"Blaine! For Pete's sake. Help me find some pictures." Even as Tracey said it, she chuckled. She knew Blaine wasn't really as arrogant as he sounded. She knew that he was loving this summer of studying science face to face, and she knew that he wanted to keep going. There was more to learn and more to explore.

The two worked together to find three good pictures in the archive app: one for light, one for color, and one for lenses. They then used the high-resolution cameras on their phones to snap mirror pictures since they were sitting next to so many.

"That's a wrap," Tracey said, satisfied. "Now, what does LINLOC say? Where will we be zipping to next? What will we be learning about?"

Now focused on the task at hand, Blaine was the first to get the information-giving application opened on his phone. "Spain!" he exclaimed. "Longitude 41° 27' 31.1" N, latitude 1° 58' 0.3" E, where we will be studying the topics of waves, sound, wave behavior, and musical instruments."

"All with the help of a local expert named Esperanza Tambor." Tracey added.

In a matter of seconds, the twelve-year-olds were harnessed, helmeted, and calibrated up. They zipped through swirls of thrilling light from Greenland to Barcelona, Spain, landing with a jerk. They were filled with nervous excitement because they never knew exactly what to expect in a new place. What would they see and experience when their post-zip sight and strength returned?

A huge ear-splitting boom and torrents of pelting water clued them in. They had landed in Barcelona in the middle of a thunderstorm!

"C'mon, Trace! We've gotta find some shelter!" Blaine shouted, helping his sister up.

They were both already drenched. It was extremely dark, so they only got glimpses of their surrounding when lightning flashed. Still, the twins moved forward, hoping to find a place to get out of the rain.

The Sassafrases figured they'd be landing at night, but in their minds, Barcelona was going to be a well-lit, bustling city. However, that currently did not seem to be the case.

There were streets, sidewalks, and buildings, but there were no people. There was no life or light. There was no electricity, other than the lightning bolts streaking through the sky. There also didn't seem to be any awnings, open doors, or shelter of any kind they could duck under to get out of the storm. The twins left their helmets and harnesses on as they looked for a solution.

They had only been here for a minute, but in that short minute, Blaine and Tracey were already feeling extremely vulnerable and far from any semblance of home. They immediately began to question themselves and the invisible zip lines. Had they spun the rings of their specially designed carabiners to the wrong longitude or latitude coordinates? Was there a glitch in the system? The zip lines were designed to drop them as close to their local experts as possible without anyone noticing. If so, then what was this place? This place wasn't close to anyone or anything.

The twelve-year-olds rushed around a couple substantially boarded up buildings and then out into an abandoned road, littered with discarded debris. They then went from soaking street, to puddled sidewalk, and down onto some railroad tracks.

Lightning flashed. Thunder boomed. Sheets of rain continued to fall.

"Look!" Tracey said, pointing. "I think there's an open door at the front of that building!"

"I think you're right!" Blaine responded in a volume louder than the storm.

The brother and sister hopped over the metal tracks, climbed up onto a cement platform, and raced toward the large building, straight toward what they thought looked like an open door. When they reached the spot, they were equally encouraged and discouraged. The doorway was indeed open, but it was because it was broken. An unkempt doorway that led into creepy darkness.

The twins paused with tinges of fear. They looked at each other in the brilliance of a lightning strike with traces of trepidation on their faces. Really though, they knew they didn't have a choice. For safety from the intense storm, they had to go through this doorway. They entered side by side. The darkness inside this place was even darker than the darkness outside the place. However, it was dry. Well, at least dry-ish. They could still hear water dripping and they were slapped in the nostrils by an established mustiness, but they were no longer being pummeled by the driving angry rain.

Will our eyes eventually adjust to this darkness? Are we truly alone, or is there someone lingering here? All these questions and a few more raced through the shivering twelve-year-olds' minds as they stood at the threshold of scary and potentially scarier.

Remaining side by side, the Sassafras twins mustered their courage and took multiple steps deeper into the building. Their eyes slowly adjusted. They could tell they were in a large open room, but the space seemed to be littered with a variety of obstacles like broken chairs and benches surrounding rows of spray-painted pillars. The twins spotted an old desk or two within the scattered and strewn items. The twins couldn't tell for sure if they were alone or not, but it seemed like they were the largest living things among all the debris.

The siblings made their way across the open space, keeping their helmets on for safety as they explored. They stepped around any and all obstacles, heading toward what looked like a wall of interior windows and doors. They weren't really sure where exactly they were going, but even though they were now inside, they were still feeling a need for shelter. This building had them out of the rain, but they still felt vulnerable in the big open space. Blaine and Tracey reached the wall and found that the windows and doors were actually the fronts of what looked like several different ticket booths. Could one of the booths give them the sanctuary they were seeking?

Blaine warily reached out and grabbed the handle of one of the doors. He moved his hand in a twisting motion. The door was unlocked. It opened slowly with a loud, yawning creak. It was dark inside, but with their eyes now fully adjusted, the twins could see the small space was empty. Here they stood at another threshold, soaking wet and cold. They were tired and weary from globe-trotting and lack of sleep. They were alone and confused as to why they had landed here.

The twins walked into the booth and leaned up against the wall. They needed to rest a minute. The darkness of the small, dark, musty ticket booth soon enveloped them. The twins closed their eyes and gave into exhaustion. They could decide what to do next in the morning.

Light spilled on the two sleeping twelve-year-olds, but it was the sound that woke them up.

"I want the *guiri* out of here!"

"Oh, c'mon, let them stay. They look harmless enough."

"*Guiri* can't be trusted. Besides, we don't even know where they came from."

"They probably came in here last night because of the storm. I'm telling you, that storm was crazy last night. I bet they came into our station to get out of the rain, you know, to get out of the thunder and the lightning!"

"Lightning? Thunder?" Tracey mumbled groggily as she attempted to blink open her eyes. She took in the old papers and ticket stubs on the floor. The dust-covered counter before them and the open door to their left. She realized she still had her helmet and harness on when she turned to find the source of the words that woke her.

"Boom! Loud noise go boom!" Blaine shouted like some kind of caveman waking from slumber.

Tracey whacked Blaine's leg, forcing him to fully open his eyes and look toward the open door. There were two girls standing over them, looking at them.

Blaine and Tracey sat up. They had fallen asleep in what had looked like a ticket booth last night. This morning, they saw it was indeed a dirty, cramped, abandoned ticket booth.

The two girls who were staring at them were standing in front of the small doorway. When a third girl approached, the two moved out of the way to let the third into the booth. All three of the girls had dark hair and resonant brown eyes. The two were probably about the same age as the Sassafrases, but the third looked a bit younger. Their faces screamed confidence, but their clothing spoke of desperation. It was a little unnerving.

"Boom? Loud noise go boom?" Blaine, for some reason, said again, this time in question form. He still sounded like a caveman.

"Sonic boom," the younger girl said. "Thunder is a natural sonic boom. A sonic boom is a release of energy and is a type of shock wave created when an object travels through air faster than the speed of sound. This causes a sudden change in air pressure that is perceived by the ear as a loud noise. Essentially, it's a compressed wave front of sound energy formed by the build-up of air molecules pushed aside by the supersonic object. Supersonic jets that break the sound barrier create this phenomenon just like thunder does. Boom."

The Sassafras' eyelids were all the way wide open. This had to be their local expert, Esperanza Tambor, but she was just a kid! And not just a kid but a little kid. Younger than twelve, for sure. And why was she already sharing science with them? Usually there was more context from within which their local experts relayed the scientific information. This girl had just walked up and started sharing. The Sassafrases were thankful for the information, but this all seemed a little strange to them. The other girls, however, were acting like this was totally normal.

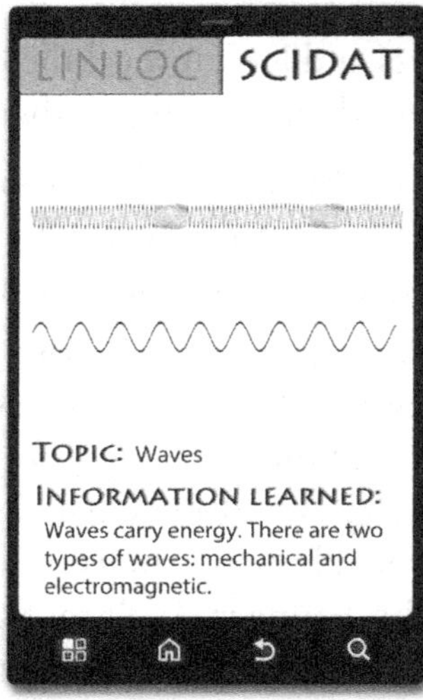

"Waves carry energy, and there are two types of waves: mechanical and electromagnetic," Esperanza Tambor continued. "Mechanical waves include both water and sound waves. They are vibrations in a solid, liquid, or gas. A shock wave, such as a thunder-clap, is a mechanical wave. Electromagnetic waves include light and radio waves. They are a different kind of vibration. Heating up food in a microwave is an example of an electromagnetic wave."

Tracey was feeling like she needed to stand up. Blaine was feeling like he needed a microwavable snack, but neither voiced these desires. Instead, they continued to listen as Esperanza continued to speak.

"Waves travel through mediums such as air, water, and glass. The different mediums affect how quickly and how far the wave travels. Waves can be described as traveling traverse or longitudinal. Traverse waves vibrate at right angles. Longitudinal waves vibrate in the same direction."

Tracey's legs ached. Blaine's stomach grumbled.

"Waves are measured by either frequency or amplitude. Frequency is the number of repeating wave patterns in a given amount of time. Frequency affects the pitch of sound we hear in a measurement called Hertz. Amplitude is the maximum disturbance in a wave cycle, or the strength of the vibration. The higher the amplitude, the more energy the wave carries." The young girl paused.

The twins thought she might be finished with her information about the topic of waves. Tracey went ahead and stood up. Blaine joined her. Now that they were all the way upright, they could see more than just the doorway area of their would-be refuge.

They could now see over the ticket counter and through the window out of the booth into the wide-open space of the building they had taken shelter in the night before.

Secret Sounds

Sunlight was streaming in from high windows, filling the humongous area with illumination. However, it still looked dirty, dusty, and forgotten. It was indeed blemished with a variety of dilapidated and seemingly discarded obstacles, large and small. The thing taking most of the Sassafrases attention was not the sunlit building but rather the people. There weren't just three girls looking at them; there were a dozen or so people staring. They were children—a few boys, but mostly girls. There was only one who was taller than Blaine and Tracey. She might be a teenager, but the rest were all young Spanish children.

"What is this place?" The twins thought and almost asked out loud. But Esperanza started sharing science again before they could.

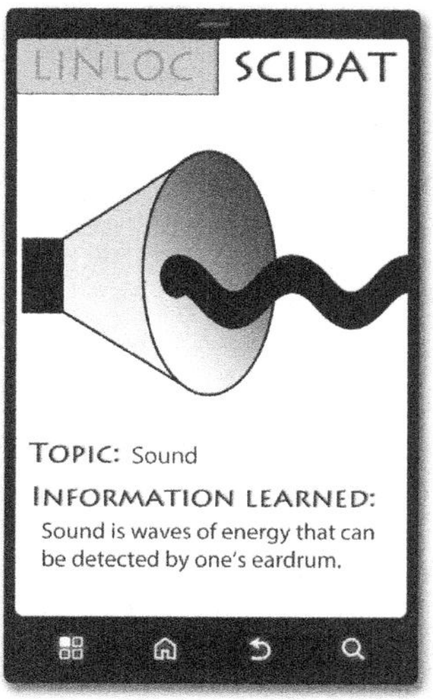

"Sound is waves of energy that can be detected by one's eardrum. Sound waves travel by vibrating the matter they pass through, which means sound can't travel in a vacuum. In other words, sound waves travel in space because they need something to vibrate. Sound waves travel longitudinally, which means the particles vibrate in the same direction as the wave travels. Sound waves move more quickly through a solid than a liquid, and more quickly through a liquid than a gas."

The Sassafras twins had thought Esperanza was finished with her science-filled dialogue. But as they listened, they could understand how her train of thought had continued, connecting the information about sound.

"A vibrating surface pushes and pulls at the particles surrounding it," the small girl continued. "These particles then push and pull against the next layer of particles, causing the vibrations to travel in the medium. Here are two examples: a speaker has a cone that vibrates, sending energy into the air, causing the particles to press together or spread out, creating sound we can hear. The human ear contains an eardrum that vibrates and can hear sounds between twenty and twenty-thousand Hertz. Anything above that is called ultrasound. Anything below that is known as infrasound."

Not one of the kids who'd gathered was talking, fidgeting, or tuning in elsewhere. They were all listening, with great interest, to what Esperanza shared.

The brilliant, well-spoken small girl continued. "And finally, echoes are sound waves that have bounced off a surface and are heard just after the original sound. Echoes can be used as locators. Ships use echoes to help them measure water depth, employing a tool called sonar. Bats and dolphins use sonar, or echolocation, as well."

In a burst of joyful spontaneity, Blaine leaned over the counter where one of the ticket booth windows had been slid open. He cupped his hands around his mouth and shouted out into the large space: "Echo!"

Not only did the sound of Blaine's voice echo, the words came back in spooky layers. He wasn't sure why, so he just grinned widely. Tracey responded with nervous laughter. All the other children stared at him suspiciously.

Esperanza looked at both Blaine and Tracey for a long second, with neither a smile nor a frown. The small girl then turned and left the ticket booth. Not only that, she kept going. She walked across the entirety of the larger room and disappeared around a corner. The Sassafras twins wanted to follow her. They at least wanted to step out of the small room they were in, but the two girls who had been talking before Esperanza started sharing about sound came back together in front of the doorway, effectively blocking Blaine and Tracey in.

"I want them out of here," the shorter of the two said.

"Oh, c'mon, let them stay," the other responded. "You just saw how they listened to Esperanza. They appreciated her and what she had to say, like the rest of us."

"But they are not like us. Look at their fancy clothes and backpacks and helmets. They are not like us."

Blaine and Tracey looked down at themselves. Were their clothes fancy? They were actually feeling disheveled and dirty after spending the night sleeping on the floor of this musty old ticket booth. Then the twins looked back up at the two girls and noticed them in a different light. The girls were pretty, but they were unkempt. Their hair needed brushing. Their clothes needed washing. They had dirt under their fingernails.

The Sassafrases now looked out through the window at the other children. All the boys and girls here seemed to be in the same state. What was this place? Why were there no parents or adults around?

"See, they are judging us. I can see it in their eyes," the shorter, skeptical girl said. She had a scowl on her face that seemed to match the holes she had in the knees of her jeans and the elbows of her long-sleeve shirt well.

The slightly taller girl, who was wearing what probably used to be a lovely dress but was now dusty and ragged, replied to her counterpart with positivity and a smile. "Alejandra, they aren't judging us. They are just trying to figure out what's going on. Our station is a new place for them. Don't you remember what it was like when you first got here?"

"No, Maya. You're wrong. They are judging us. They are *guiris*. Foreigners. They can't be trusted. And I want them out of here!"

"Whoa! Whoa! Whoa," the teenage-looking girl interrupted as she approached the doorway of the booth. "Settle down, Alejandra. Maybe these two can be trusted and maybe they can't, but you and Maya haven't really even given them a chance to speak yet, so how would we know?"

The two girls who'd been conversing and blocking the doorway scooted to the side and let the older girl in. She was naturally beautiful, but like the others, she was dirty and tousled. She was most certainly a teenager. There was wisdom in her gaze that made her seem even older. She looked directly at the Sassafrases and like Esperanza earlier, she was neither smiling nor frowning.

"Who are you? What are your names?" she asked, no emotion showing on her face.

"We are the Sassafrases," Blaine said. "Blaine and Tracey Sassafras. We came here to Barcelona to learn about science and physics. This is . . . Barcelona . . . isn't it?"

"Of course it's Barcelona, Blaine," the girls answered. "Just not the pretty part."

She took another long look at the American siblings. Was that a slight smile on her face? "Blaine and Tracey Sassafras. Here to study science, huh?"

Both twins nodded.

"My name is Valentina," the girl said, pointing at herself. "Thanks to Esperanza, I have grown to appreciate science myself." She then pointed at the other two girls. "This is Alejandra and Maya. Alejandra thinks you two can't be trusted. Blaine and Tracey Sassafras, can you be trusted?"

Tracey nodded and attempted to answer. "Yes. I think . . . we can . . . for sure . . . be trusted?"

"Was that a question or an answer?" Valentina asked the

Sassafras girl.

"An answer," Tracey said now, a little more confidently. "We are trustworthy."

The teenage girl took another long look at the Sassafrases. Slowly but surely the slight smile they had seen earlier fully formed on her face. "Blaine and Tracey Sassafras, welcome to our station. Come out of this booth and meet everyone else." Both twins shimmied out of their harnesses and their helmets, storing them in their backpacks before they headed for the exit.

Valentina led the twins out of the ticket booth, through the doorway, between the smiling Maya and the still scowling Alejandra, out into the large open room. "*Los Granujos*, meet Blaine and Tracey. Blaine and Tracey, meet *Los Granujos*."

"*Los Granujos*?" Tracey asked.

"It's slang for ragamuffins in Spanish," Valentina said. The teenage girl swept her arm over the group. The twins saw a girl with jet-black hair, one with curly hair, one wearing a hat, and one staring at them with brilliant green eyes. They saw a boy wearing glasses, one that looked like a body builder, and one with a scar on his forehead. They also noticed a little boy who constantly wiped his nose.

Both twins lifted their hands and waved. The children waved back.

"The Ragamuffins?" Blaine now asked. "Why are you all called that?'"

"Because we are all ragamuffins," Valentina answered matter-of-factly, including herself in the group. "All of us have been forgotten or left behind. We've got no parents, no guardians, no teachers. So we take care of each other. We gather and stay hidden here at our station."

"Your station?" Tracey asked.

"Yes, our station," Valentina confirmed. "This place is an old railway station. It's been abandoned for a long time. It used to be an important part of Barcelona, but not anymore. It's outside the city center. It's been forgotten and left behind, just like us."

"You . . . live here?" Blaine's head swiveled around, taking in the large space which looked nothing like a home.

"Yes, Blaine," the teenage girl answered like she'd already been very obvious about this fact. "We live here. This isn't the only space in our station. This building is huge. It's like our very own broken-down mansion. There are places for all of us: nooks, crannies, rooms, and booths we can hide in. We live here, and now you can, too."

"Well . . . I don't know if we are . . . quite ready to move in . . ." Blaine bumbled.

Tracey was also a little taken aback by Valentina's sudden invitation but deeper than that, she felt sympathy for this teenager and these children called *Los Granujos*. "I'm sorry that you're all here alone without any family," Tracey said with authentic compassion, knowing her and Blaine couldn't truly understand. They had a mom and a dad to go home to when they finished their summer of science-learning.

Valentina had a look on her face that showed she appreciated Tracey's tenderhearted words. The other children all seemed to share her sentiment. "Don't worry, Sassafrases," Valentina said. "We don't need families because we are each other's family."

"But what about Esperanza?" Blaine asked. "Is she a part of this family? Is she a part of *Los Granujos*? Where did she go?"

"Blaine, bro, *hermano*. What's up with all the questions with obvious answers," Valentina said to Blaine like they'd known each other their whole lives. "Of course, Esperanza is a part of the family. But she's special. She's—"

The teenager was cut off by a sudden shout from one of the

kids who was over closer to the exterior door the Sassafrases had entered through last night.

"Somebody's coming! An adult! Everyone hide!"

CHAPTER 11: THE RAGAMUFFIN ORCHESTRA

Musical Memories

"*Los Granujos*! It's just me. You can all come out!" A friendly female adult voice called out.

Blaine and Tracey looked at Valentina who was here with them in the small ticket booth the Sassafrases had slept in last night. The twins were looking at Valentina now to confirm with this teenage leader of *Los Granujos* if it was indeed okay to come out of hiding like the woman had just implored them to do.

Valentina smiled. "It's just Maria. We're good."

The teenager stood up and walked out of the booth into the large room. The Sassafrases followed her. The other children emerged from their hiding spots and approached the woman who had entered their station. Maria had a kind look about her that matched her voice. She was carrying an armload of large round loaves of bread, which she immediately handed out with a smile to

Los Granujos as they surrounded her.

"Whoa, whoa, whoa," Valentina scolded but not harshly. "Slow down and use some manners. We're poor but we're not barbarians. Give Maria a little space and make sure to thank her."

Multiple iterations of *gracias* were offered up as grimy hands grabbed crunchy bread. "I come bearing bread, but I also come bearing news," said the Spanish woman. She looked like she could've been a grown-up version of many of the girls here, albeit a little cleaner, a little less dingy, grown-up version.

"Probably bad news," Alejandra said pessimistically.

She was immediately countered by Maya. "Oh, I bet Maria has wonderful news for us. News of hope and opportunity."

"Well, it's kind of a little of both," Maria said, now completely bread free. "The concert is happening tonight!"

"Tonight?" Many of the children responded aghast with open mouths full of half-eaten bread.

"Tonight," Maria confirmed. "Maestro Martinez made the announcement yesterday. He said the orchestra is ready. He still doesn't know the song was stolen. He doesn't know it's Esperanza's. He's going to give the credit to that other woman!"

Valentina now looked very defensive and serious. "Can we get in tonight? Will you help us?" She asked the woman.

"Of course," Maria responded. "Are your instruments ready? Can you play her song perfectly?"

Now Valentina looked unsure but still serious. "I'm going to give you a solid maybe on both of those questions."

"Maybe?" Maria said, now looking unsure herself.

Valentina nodded.

Maria looked at Valentina and then around at all the orphaned and abandoned children. Her sympathy for them was evident.

"Tonight is your shot," she said. "It may be the only chance you have to get Esperanza the credit she deserves. Of course, I will help you tonight, but I just hope you all are ready."

"I hope we are ready, too," Valentina responded in earnest.

Maria stayed and chatted with the children a little while longer before she had to go. *Los Granujos* bid her *hasta luego* and then they turned their focus inward. They all seemed very clear-minded and determined about doing whatever it was they were going to do tonight at this concert.

The Sassafrases were also determined, but they were utterly confused about what was going on. The twelve-year-old *guiris* turned toward Valentina and were about to bombard her with their inquiries.

Sensing it was coming, the teenager beat them to the punch. "I know, I know, Sassafrases. You are wondering what in the world is going on."

Blaine and Tracey nodded in confirmation.

"Esperanza is one of us, but she is different. Always has been," Valentina said, pointing at herself and all the other children. "The rest of us kind of wandered in here, found our station, and found each other. But not Esperanza Tambor. She was left out in front of our Station, in the street, inside a big bass drum or, in Spanish, a *tambor*. That's why we call her Esperanza Tambor. We don't think she was forgotten. We just think she was accidentally lost. I'm the one who found her out there in the drum in the street. She was probably only five or six-years-old at the time. She didn't remember anything beyond her first name."

The teenager paused with a far-off look on her face that revealed she was reliving the memory. "Esperanza was different in other ways too," Valentina continued. "She's a true genius! She was already speaking at least three languages when we found her, and then we all started to notice she could remember anything she read."

Valentina laughed out loud at this memory, as did some of the other *Los Granujos* who were listening to her reminisce. "It kind of became a game for us," the teenager continued with a smile. "We go out searching in the streets, in the garbage, or over in the dump for anything Esperanza could read: books, magazines, food labels, anything in any language. We would bring all of it back to our station and give it to her to see if she could read it and remember it. She could! Unlike her mysterious past, she could remember everything we gave her to read, without fail. It was absolutely amazing!"

Los Granujos were all nodding their heads in agreement.

"And not only could she read it," Valentina said with bright eyes. "She could also teach it! That's how we are even conversing with you *guiris* right now. Because Esperanza taught English to all of us!"

"*Et Fraincois*," one of the girls added.

"*Und Deutsch*," one of the boys piped in.

"We gave Esperanza a room to herself around that corner and up a set of stairs," Valentina pointed at the corner the twins had seen Tambor disappear around earlier. "We helped her fill that room up with words. You know, any words that seemed to have worth, especially books. Esperanza really seemed to love science books. She would teach us everything she was reading and learning about the whys and the hows. Many of us have grown in both our appreciation and understanding of science because of this."

There were several nods bobbing amidst *Los Granujos*. Valentina had been talking pretty quickly, but now she slowed her speech down as if the subject she was bringing up deserved reverence.

"Then one fateful day, I found a music book. It was a big, thick thing, filled from cover to cover with concertos, sonatas, and symphonies. Words had fed Esperanza's brain, but the musical notes fed her soul. Not only did she quickly learn to read the language of music, she also almost as quickly began to write it. Esperanza began composing song after song, some with lyrics and some without. She began filling up her room with the songs she was writing. It was all we could do to find her enough paper and writing utensils in the garbage or at the dump to keep her going."

Valentina laughed out loud even now as she remembered it. The hard-shelled, soft-hearted teenager continued, "The only problem with all the music Esperanza was writing was that we had no way to play it. There was no doubt in any of our minds that what she was writing would be absolutely lovely, but how would we ever be able to hear it without any instruments?"

The Sassafras twins nodded.

"So you know what we started to do?" Valentina asked.

Blaine and Tracey raised their eyebrows in an unspoken question.

"We started to travel across town at night and sneak into *Palau de la Musica Catalana* for the evening performances."

"*Palau de la Musica*?" Tracey asked.

"*Catalana*?' Blaine added.

"Yes," Valentina confirmed, like the twins should know the place. "It's only the finest and fanciest concert hall in all of Barcelona."

The twins nodded as if they knew what she was talking about.

"Maybe we couldn't hear Esperanza's music," Valentina continued. "But at least we could hear the wonderful music at the Palau. We especially love the orchestra performances conducted by Maestro Martinez. Esperanza taught us all about the musical instruments we saw playing there. She shared about how they are often grouped by how they make sound. For instance, stringed instruments make sound when strings vibrate either by being hit, plucked, or slid over by a bow. Violins and cellos are stringed instruments and so are harps."

Maya interrupted. "She also taught us that musical instruments work by making sound waves. And the size, shape, and type of material all affect the sound that is produced. So big instruments like the bass have a deep sound, while little instruments like the piccolo have a high-pitched sound. And wind instruments, like the piccolo and trumpet, make sound when a column of air inside them vibrates when the player blows through a mouthpiece."

Sniff. The little boy wiped his nose and said, "Esperanza taught us that percussion instruments make sound when they are beaten, scraped, or shaken."

The girl wearing the hat spoke next, "Esperanza even explained how many instruments, like a violin, for instance, have a sound box that resonates to create richer sound. Whereas electric instruments use an electric amplifier for sound instead of a sound box. We don't have any electric instruments here at the station, mainly because we don't really have access to electricity. But they are pretty cool instruments because they can control amplification as well as add special effects like echoes."

Blaine started to cup his hands around his mouth to try to make another echo, but before they even got halfway to his mouth, he was interrupted by the teenage drummer.

"Don't even think about it, Blaine," Valentina snapped like an older sister. The Sassafras boy smiled sheepishly and dropped his hands.

"The last thing Esperanza taught us was about pitch. Pitch is the highness or lowness of sound. Sound with high frequency has a high pitch. Sound with low frequency has a low pitch. Typically, large instruments like the cello or the bass drum have a lower pitch, while smaller instruments like the clarinet or the violin produce a higher pitch," the girl with the green eyes shared.

Heads were nodding all around the station upon this statement.

"One particular night, we got caught sneaking into the *Palau de la Musica Catalana*." It seemed to the Sassafrases like Valentina should've been frowning when she said this, but instead she was laughing again.

"Good thing it was Maria that caught us," Maya interjected, laughing herself.

"Yes, good thing," Valentina agreed.

"You mean the Maria we just met?" Tracey asked.

"That's right," Valentina confirmed. "Maria Aldana, the one and only."

"She wasn't mad that you guys were sneaking in?" Blaine asked.

"No, Blaine, she wasn't mad," the teenage leader of *Los Granujos* answered, continuing to talk to the Sassafrases like they'd known each other their whole lives. "Maria is a janitor at the concert hall. She used to be an urchin like us until she started working at the concert hall. She loved the fact that we appreciated the fine music because she appreciated it so much too. Evening trips to the *Palau de la Musica Catalana* became our most consistent outing. Maria found a great little hidden spot upstairs behind the organ pipes where we could all fit comfortably and listen. We grew in our knowledge and appreciation of the art of music."

This made the Sassafrases smile.

"Yes, we appreciated it, but we still couldn't play it." Valentina communicated the frustration she felt. "We watched all those musicians playing their instruments, and we just knew we could do it if we had a chance, if we had instruments of our own. We just knew we could play all the songs they were playing, but even better than that, we could play all of Esperanza's songs if we had instruments."

All at once, the Spanish teenager furrowed her brow in what looked like anger. "That's when she showed up," she said. "That woman!"

"Sí! The Woman!" The ever-negative Alejandra cut in.

She pointed two thumbs down and then literally spit on the ground in disgust. Several other children followed suit.

"I don't know how she found us or how she knew about our station or about Esperanza's music, but she found us. Maybe she saw us sneaking into the concert hall. Maybe she followed us. I don't know," Valentina said, brow still furrowed. "But she came in acting all happy, and she had instruments, lots of instruments. She promised to help us—to train us and teach us how to play all the instruments. She even made it sound like, if we were good enough, we could maybe play the Palau someday."

Alejandra cut Valentina off with another spit on the ground, then said, "But she was a liar. That woman was a liar!"

"And not just a liar, but also a cheat and a thief," the usually positive Maya added with a frown on her face.

"And a meanie," sniffled the little boy with the perpetually runny nose.

"Yes, The Woman was definitely a meanie," Valentina agreed. "She was nice at first, but then slowly but surely, she became more rigid. She turned cruel. She made us practice long hours. She yelled at us if we didn't play the music perfectly. She threatened us and even mocked us for being orphans. She prohibited Maria from visiting us. It was just practice, practice, practice around the clock, with no encouragement, not enough breaks, and not enough food or water."

The twins stood there with mouths open. They were starting to understand why these children did not trust outsiders.

Valentina paused to let out a sigh that sounded a little like a growl. "The Woman was especially cruel to Esperanza. She pushed her harder than any of the rest of us. She wasn't just making Esperanza practice her instruments, she was also making her write songs. I don't think The Woman even let Esperanza sleep."

"At least not until Esperanza wrote The Song," Maya interjected.

"That's right," Valentina confirmed. "One day we all sat down to start what we thought would be just another excruciating day of practice with The Woman. She directed us to start playing a new song Esperanza had written the night before."

Valentina again slowed her delivery and began speaking reverently. "From the first note of the first line, we all immediately knew this song was special. What we heard coming from the breath and strings of our instruments was the most beautiful sounding thing any of us had ever had the privilege of listening to. It was even more poignant than anything we'd ever heard the orchestra play at the *Palau de la Musica Catalana*."

Every single one of *Los Granujos* nodded in agreement. Some even had tears in their eyes. Valentina was one of those, and as the teenager continued, a couple tears spilled over the bottom lids of her eyes, tracking across her dusty cheeks.

"Esperanza's song enraptured us. Every note spoke to our souls. Every stanza embraced us. It was such a beautiful song, The Woman herself even stopped in her tracks. After we played the prolific song just one time, The Woman let us stop practice for the day. She packed up all the instruments, took the sheet music, and left. She didn't yell, scream, or shout. She didn't say goodbye or *adiós*. The Woman didn't say a word. She just left."

Valentina paused, leaving Blaine and Tracey wondering if that was the end or if there was more. There was more.

"The next day, The Woman did not return to our station," the teenager continued. "Nor the day after, or the day after that. We didn't know why she'd shown up in the first place, and now we were just as confused by her departure. We didn't know what she wanted from us or why she was so insistent on teaching us to play the instruments. Was there a purpose or a plan?"

The Sassafras twins shrugged.

"We were happy to be free from her control, but we still worried," the older girl explained. "Was she going to come back? Or was she gone forever? We didn't miss her, but we sure did miss those instruments. Esperanza's song had moved us all. It was the most wonderful thing our ears had ever heard. Would we ever hear it again? Would any of us ever get to play a violin, cello, trumpet, clarinet, or any other instrument ever again?"

The twins shrugged once more.

"We eventually found out what happened," Valentina said angrily with a scowl back on her face. "After a whole week of us wondering what happened, Maria showed back up at our station. She told us The Woman had gone to *Palau de la Musica Catalana* and had procured a meeting with Maestro Martinez. In this meeting, she presented Esperanza's song to the Maestro, but instead of giving credit to Esperanza, The Woman told the Maestro she had written it! She stole our girl's song and claimed it as her own!"

Now the Sassafrases gasped involuntarily.

"Maria said when Maestro Martinez had the orchestra play the song and he heard if for the first time, he was absolutely speechless. The Woman peppered the Maestro with questions after the song was played. Would he let the orchestra play this song in front of a paying crowd at the Palau? Would he give her the credit she deserved for writing the song? Did he think she would become famous? Maria said the Maestro remained speechless for a long time, but eventually replied, telling The Woman that she needed to be patient and that he would be in touch. That was a few months ago now."

The twins now both had scowls on their brows. How could someone be so cruel and steal a song from an orphan like this?

"So, did Maestro Martinez ever get back to The Woman?" Blaine asked. "Did he ever say whether or not his orchestra would play the song?"

Valentina responded to the Sassafras boy with an eyeroll, like an older sister. "That's what Maria was telling us about earlier. The Maestro has confirmed that he will indeed conduct his orchestra to perform the song in a concert tonight at the *Palau de la Musica Catalana*. And he's going to give The Woman the credit for composing it!"

"Well, that's just not right," Blaine responded, still scowling.

"We know, Blaine. It's not right, but we've got a plan." Valentina grinned.

"What kind of plan?"

"A good plan, Blaine. A really good plan."

"Well, what is it? What's the plan?" Blaine asked impatiently, also treating Valentina like a sibling.

"Esperanza helped us make our own instruments," the teenager said with a sly smile.

"What? How?" Now it was Tracey who asked.

"She read somewhere that it was possible to make instruments out of garbage. So that's exactly what we did. We scavenged our normal spots for the right kind of junk and materials, and we made our own instruments. They sound great! We've been practicing on our own. We've got all the instruments we need to be a pocket orchestra. We have violins, cellos, trumpets, clarinets, tubas, trombones, cymbals, drums, and even a harp! We can play Esperanza's song again!"

Blaine's brow went from furrowed scowl to scrunched question. "That's cool and all, but how is that a plan?"

"You're killing me, Blaine," Valentina puffed, pretending to be exasperated. "We are going to grab all our instruments and hike them over to the *Palau* tonight. Maria is going to let us in. When it comes time for the orchestra to play Esperanza's song, we are going to rush the stage and play it ourselves. That will prove that Esperanza wrote the song and not that Woman!"

"Ohhhh," Blaine responded, now understanding.

Tracey asked an unexpected question. "Does Esperanza's song have words?" Valentina looked thoughtful. "I don't know. Maybe. We've only practiced the music, but I've seen Esperanza silently mouth words several times while we were playing it. Also, I'm pretty sure there were words written on the copy of the music that The Woman took with her. If there are words, only Esperanza knows them. And maybe The Woman."

"Where are the instruments?" Blaine asked.

"Oh my goodness! So many questions!" Valentina laughed. "Too many for now. Let's just go find Esperanza. She wasn't down here when Maria came. Let's go tell her the news. Let's see if she wants to move forward with the plan."

Wave Behavior

The group of ragamuffins made their way across the large room and then around the corner Esperanza Tambor had taken earlier. Now they were in a short hallway. At the end of the hallway was a set of old wooden stairs. The stairs led up to a single solitary door. It was a large, heavy-looking wooden door that probably used to be considered ornate but now just looked dusty and dated. The door was closed.

Valentina paused at the bottom of the stairs. "It's a lonely set of stairs leading to a lone doorway that opens up to a room situated all by itself."

The teenager sighed. "I don't know what it was back when the train station was up and running, but now it's Esperanza's room. It's a great place for her to be alone. For her to read and study science and write music."

Every step either cracked or squeaked as the group walked up. Valentina reached the door, knocked, and then slowly opened it up.

"*Bienvenida*," the Sassafrases heard Esperanza say from inside the room.

Blaine and Tracey stepped in behind the teenager and took a quick wide look around. It was a rather large room with floor-to-ceiling bookshelves on every wall. Every shelf was full of something readable. In the middle of the room, there was a small cot, presumably where Esperanza slept, and there was a simple wooden table piled high with papers and books. The room was lit by a large circular skylight. Esperanza was standing near the table with a book in her hand, and she looked happy to see them.

The Sassafras twins were struck again by how small she was. How could someone so young with such a tough lot in life be so smart? It truly was incredible.

"I was just reading about sound wave behavior," the brilliant little girl said, and then launched into sharing more.

"When a wave hits an obstacle or travels into a different medium, its speed, shape, or direction change," Esperanza said. "The original wave, before it experiences change, is called the incident wave. When an incident wave hits an obstacle, part or all of it bounces back. The wave that bounces back is called the

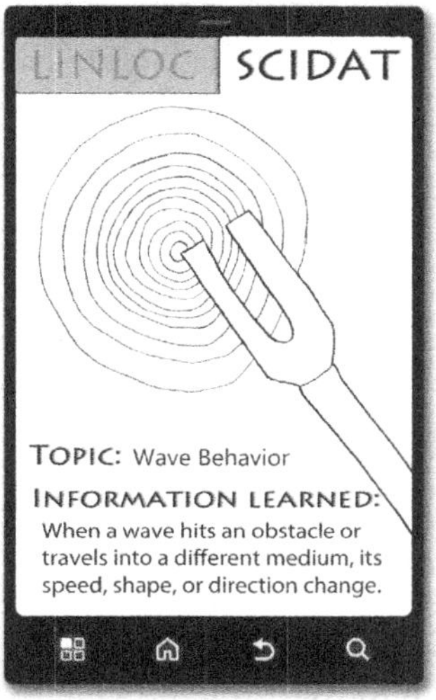

reflected wave. The angle at which it is reflective depends on the surface of the obstacle it hits. When an incident wave passes into a new medium, it can change speed and direction. The wave that changes direction is called the refracted wave."

Blaine sort of recalled his eerie attempt at an echo earlier in the day. He kept listening to learn how else waves behaved.

"When an incident wave goes through a gap in an obstacle, some of the wave passes through, making it bend and spread out. This is an example of 'diffraction,'" Esperanza continued. "And then 'absorption' is the measure of the amount of energy removed from the incident wave as it passes through the thickness of a medium."

The girl put the book she was holding down on one of the stacks on the table. She then stepped over to Blaine and reached up and put a hand on his shoulder. She looked right at him and said, "Blaine, earlier in the big room you were hoping for your incident wave to become a reflected wave. You were hoping the sound wave you sent out would bounce back as an echo. And it did just that! In a room that large, your incident wave was refracted and reflected and sent back to us in an echo."

Blaine smiled with a big toothy grin.

Esperanza took her hand off the Sassafras boy's shoulder and continued. "Loudness is measured in decibels. The higher the decibel, the louder the sound and the larger the amplitude of the sound wave. The amplitude of a sound wave decreases as it travels, which is why the volume of a sound decreases the farther away you are from a source. And then the last thing I'll say about all this is this: when two or more incident waves meet, they interfere with each other. Sometimes the waves meet and the result is larger. This is called constructive interference. Sometimes the waves meet and they diminish or cancel each other out. This is called destructive interference."

All this talk about wave behavior made Blaine want to try for another echo. The Sassafras boy cupped his hands around his mouth and was about to shout out, but before he did, Esperanza placed her little hand back on his shoulder.

"I don't believe it's going to work, Blaine," she said kindly. "The walls of books in here will just absorb the sound."

Blaine dropped his hands.

Valentina raised hers. "Okay, enough of that! We didn't come up here to make echoes. We came up here to tell Esperanza the news!"

Los Granujos and *los* Sassafrases were now in a completely different part of the station. Esperanza was directing everyone to set up for practice.

Valentina hadn't lied earlier when she'd referred to instruments being made from junk and garbage.

"Here is our string section." Esperanza pointed at some violins that looked like they were formed and shaped from rusty metal sheeting, a cello made with an old metal barrel and wood scraps, and a large harp made from a curved tree limb. All these instruments were equipped with creatively yet perfectly placed strings.

"And these must be the wind instruments," Tracey said as Blaine pointed to what looked like a group of children holding a trumpet, a trombone, a saxophone, and two clarinets. The Sassafrases were pretty sure the trumpet was made out of an old muffler from a car. The trombone looked like it was made mainly of PVC pipe. The saxophone had lots of small and moving pieces made from an assortment of things, including but not limited to spoons, forks, and metal pins or buttons. And the two clarinets incorporated rusty metal pipe and cut plastic soda bottles.

Valentina sat down to face the drum set made from trash can lids and an assortment of buckets and cans. Her drumsticks looked to be old wooden spoons. The teenager sent a sly smile the Sassafrases' way and kicked the large base drum, making a nice deep sounding thud.

The sniffly boy handed Tracey and Blaine carboard tubes, and said, "Maracas."

"Hey Sassafrases, find a spot to sit down," Valentina chirped. "Now Esperanza is going to start conducting. You two are going to be a part of our pocket orchestra with those maracas in your hands."

Valentina looked at Tambor for confirmation. Esperanza nodded. The twins took their place among the ragamuffin band. On top of making their own instruments, *Los Granujos* had also made enough chairs and music stands for everyone. Even now, the stands held the handwritten sheet music for Esperanza's song.

When Esperanza heard about the news Maria Aldona had brought earlier, she'd jumped into action. The small girl had led the way to this room and got to work.

Even now, as Esperanza Tambor, the little girl found on the street in the bass drum, stood in front of her makeshift family preparing to conduct them, a lone tear formed and then tumbled out of her eye. A tear that seemed attached to much deeper emotions.

Valentina was playing the drums. The cranky Alejandra sat at the harp and actually looked rather peaceful. The ever-optimistic Maya was at the cello. The sniffly boy had the saxophone. The boy with glasses had a trombone. Two other boys both grasped trumpets. A pair of girls were preparing to breathe life into their clarinets, while another set had bows ready to awaken their violins.

Los Granujos sat completely still and quiet, waiting for Esperanza Tambor to lead them.

Using a stick from a tree branch as her baton, she raised her hands in the air, and then like a mini maestro, her conducting began.

From the very first note, Blaine and Tracey could tell this song was special. Its melody was captivating. Its harmony was enrapturing. Also, to their surprise, the hodgepodge instruments sounded perfect. *Los Granujos* were all true musicians, and their homemade instruments were proving worthy. Then, maybe most surprisingly of all, Blaine and Tracey's maraca playing was blending well.

As the Sassafras twins shook their maracas, they couldn't help shake the feeling that there was something so familiar about this song. It was like some long-lost memory that was distant yet familiar all at the same time. The mini maestro directed them flawlessly to the very last note.

Los Granujos finished with a flourish. The song was over. The Sassafrases were speechless. No words were coming from anyone else either. Esperanza now had multiple tears on her face. It was a beautiful moment.

Finally, it was Valentina that spoke up. "That sounded pretty much perfect, Esperanza. Don't you think?"

Esperanza nodded.

"Still, do you think we should practice it a few more times?" the teenager asked.

The prodigy nodded again.

Several hours later, Blaine and Tracy found themselves

traveling with *Los Granujos* through the evening darkness from the station toward *Palau de la Musica Catalana*. They were going from the rundown part of Barcelona to what Valentina called the fancy part.

All the children, with the exception of the twins, had done this journey many times. So even though they were all carrying instruments, some of which were rather large, the going was pretty easy. In no time at all, they were in a cleaner, more manicured, well-lit part of the city.

The buildings were full instead of vacant and maintained instead of dilapidated. Scores of people were out and about, walking the sidewalks and avenues without care or fear.

Valentina, of course, was leading the group of instrument-carrying ragamuffins, and she was trying to keep out of sight. But really, it didn't seem like anyone was paying the line of children much attention. Eventually, the pseudo-stealthy traversing group rounded a corner, and there it was: The *Palau de la Musica Catalana*. It looked kind of like a castle to the Sassafrases, a castle situated right here in the tight streets of Barcelona.

The group skirted around the front of the building to the far side of it, and there standing at a cracked-open side door already waiting for them, was Maria Aldana. She waved them over, and without saying a word, the whole group filed into the building. The kindly janitor led the children through a series of winding vacant hallways and then up a set of stairs. It was clear to the twins they were using maintenance corridors that neither guests nor musicians used.

At the top of the stairs, Maria unlocked and opened a door for them. *Los Granujos* quickly shuffled into the room behind the door, still carrying all their instruments. The Sassafrases recognized this was the hidden spot upstairs behind the organ pipes Valentina had mentioned earlier.

The small room had walls, but the front wall where the organ pipes were was made of some kind of mesh, allowing them to see out into the concert hall, while at the same time hiding them from view. They could see the ornate auditorium filled with elegant ladies and fine gentlemen all dressed to the nines, talking in hushed tones, waiting for the concert to start.

"Okay, to get backstage," Maria whispered, "Take that door right there."

She was pointing to the only other door in the room, opposite from the door they'd entered. "It leads straight down to backstage. Then, from there, you can rush the stage when the time is right."

The children nodded. Maria smiled, gave a little wave goodbye, and then quietly exited.

Los Granujos remained quiet after their janitor friend's departure, yet there was a nervous energy in the room. The Sassafrases sure felt it.

"Are we really going to do this?" The twins thought. "Are we really going to rush the stage? Can we play the song perfectly like we did in practice? Will Esperanza really be proven as the true composer of her song?"

All at once, a hush fell over the gathered crowd out in the auditorium. The orchestra was taking the stage. Forty plus musicians, all dressed smartly in black, walked out and took their appropriate seats. There were wind instruments, stringed instruments, and percussion instruments. The Sassafrases were certain that whatever they played was going to sound beautiful.

Then, Maestro Martinez came out. He was a handsome older man with mid-length white curly hair and wonderful posture. He took his place front and center at the main stand facing his musicians. His back was to the audience, the tails on his tuxedo evident to all. The Maestro took up his baton and then with precise and graceful movements, he began conducting his orchestra.

Blaine and Tracey immediately felt embraced by the music. It was absolutely wonderful. They looked around at their companions as they listened. Here, they were not ragamuffins. Here at the *Palau*, they were connoisseurs.

Maestro Martinez skillfully conducted the musicians as they seamlessly transitioned to and from several different symphonies and concertos. The onlooking children could've listened forever, but then there was a pause. The music stopped and the maestro turned to face the audience. As a spotlight illuminated his handsome face, he made an announcement in Spanish.

"This is it!" Valentina whispered. "They are going to play Esperanza's song. And he's about to introduce The Woman as the composer!"

"*Esta proxima sinfonia se titula, 'Encuéntrame',*" Martinez told the audience in Spanish. "*Esta es una nueva canción que nunca antes se había tacado.*"

"What's he saying?" Tracey quietly, but urgently, asked Valentina.

"He said they're about to do a song called 'Find Me,' a new song that's never been performed anywhere before," the teenager answered.

"*Y la compositora de esta impresionante sinfonía es . . .*"

A spotlight now shone down on the front row of the audience, highlighting a certain woman.

"The Woman!" Valentina gasped.

The news was good. It made him smile. He was finished stacking bricks and slinging mortar. From the outside, the Forget-O-Nator was now securely situated in what looked like an official extension of the junior high building. From the inside, it was connected to a hallway by a normal-looking classroom door. It was a perfect memory-wiping trap.

The stage was set. The invitations had been sent out. The junior high reunion was a go. It was so close now, he could almost feel it: the end. The end of everyone's memories.

CHAPTER 12: CAPE TOWN THE WHOLE NIGHT THROUGH

Electrical Energy

While The Woman basked in the glory of receiving credit for composing a song she didn't write, *Los Granujos* grabbed their handmade instruments, exited their hiding spot, went down the stairs, and arrived backstage. Through thick curtains, they heard the orchestra playing Esperanza Tambor's song.

"Are you ready?" Valentina asked Esperanza.

The brave little girl nodded. She then tugged at a gap in the curtains and rushed out onto the stage, her friends following her. The orchestra and its conductor flinched and stopped playing. Little Esperanza took her place at the main stand. *Los Granujos* muscled into different seats with their instruments.

After only a slight pause and a deep breath, Esperanza lifted her small tree branch baton and led her ragtag band of ragamuffins.

The members of the orchestra gasped. Maestro Martinez gasped. The crowd gasped. As the music reverberated from the trash-made instruments, it was immediately clear these children knew how to play.

Who were these children? How did they know this breathtaking song if it had never been performed before? Who was the tiny conductor? The Sassafrases could almost ask the crowd's questions for them as the two shook their cardboard maracas.

The makeshift pocket orchestra played Esperanza's song perfectly, but then something happened that no one had been expecting, not even *Los Granujos*. After the final note, instead of bringing the song to an end, Esperanza directed her band to keep playing and to continue again from the beginning, which they seamlessly did, but this time, she grabbed a microphone and turned around to face the audience.

While the eight girls and five boys continued to play their instruments, Esperanza Tambor sang. Her song had words! And her voice was rich, emotional, and resonant.

As Esperanza sang the last word, *Los Granujos* played the last note. The *Palau de la Music Catalana* fell silent. The soft sound of crying broke the quiet. Esperanza's eyes and cheeks were streaming with tears.

All at once, in overwhelming delight, the entire audience stood to their feet in uproarious applause. The orchestra joined them, as did the Maestro. The cheers and applause for Esperanza and her pocket orchestra continued, with no sign of stopping.

Tracey again leaned over to Valentina to ask the teenager for translation.

"What were the words to her song? What did Esperanza sing about?" The Sassafras girl asked Valentina, who seemed to be caught up in the moment like everyone else with her own tears on her cheeks. She was, however, willing to grace Tracey with an answer.

"The song is about searching for her mother. For Esperanza, this was never about getting any recognition," Valentina said, with tears in her eyes. "This was always about finding her mom. She must have remembered more than we thought."

Later that night, up in the hidden room behind the organ pipes, the Sassafras twins found themselves alone with full hearts and phones. Their phones were full of SCIDAT data, all of which they had just sent to the data screen in Uncle Cecil's basement. Their hearts were full because Esperanza Tambor had received the credit for composing her song, and *Los Granujos* had been lauded for their musical performance. But their heads were full of lingering open-ended questions they doubted would ever get answered.

The standing ovation Esperanza and the young pocket orchestra received for performing "Find Me" had lasted for a long time. Maestro Martinez announced that the children's song had been the glorious finale, and the concert had concluded for the evening. A long line of compassionate, impressed, and wealthy patrons and concert attendees waited to talk to the children. Many offered to help in any way they could, both to find Esperanza's mother and also to support and care for the rest of the young musicians.

After the crowd left, the Maestro made three more announcements. First, Maria Aldana, the janitor, was getting a raise. The children had given her credit for enhancing their love and knowledge of music by allowing them access to the *Palau*. Second, he would use all the resources he could muster to help find Esperanza's mother. Third, he vowed to prosecute The Woman to the full extent of what the law would allow.

Blaine and Tracey wondered if the song would help Esperanza be reunited with her mother. But they were content with how things were changing for *Los Granujos*. Barcelona had been a wonderful location. They had learned so much and made some great friends.

"Cape Town, South Africa!" Tracey said as she looked at their next learning location on the LINLOC application on her phone. "Longitude 33° 48' 44.9" S, latitude 18° 29' 3.7" E. And we'll be studying the topics of electricity, circuits, batteries, and electronics."

"The local expert helping is somebody named Lubanzi LeRoux." Blaine added, reading from his own screen.

The siblings, who were already harnessed and helmeted, calibrated their specially designed three-ringed carabiners to the coordinates that would zip them from Barcelona to Cape Town. Seconds later, they were zipping on invisible lines through swirly light, full speed ahead.

Quickly, the zip-line travel came to a stop, their carabiners automatically unclipped. They slumped down momentarily without physical strength or sight. Before Blaine and Tracey saw anything, they heard, a voice.

"Oh, there you two are. Great! I'm so glad you're here. How are your voices sounding?" It was a deep, authoritative yet calming male voice.

The twins blinked their eyes and righted their slumping bodies as sight went from blindness to white to full color, and strength went from numbness to lethargy to full capacity.

"Our voices?" Blaine asked the tall, middle-aged man in front of them. He sported warm, casual wear topped off with a head full of short, braided microlocs.

"Yes, of course, your voices," the dark-skinned man laughed.

The Sassafras twins did not know what he was talking about. Where were they? They looked around their new setting. It looked like a small conference room or office. The man was standing on the opposite side of the room next to a door that looked like it exited to a hallway. They were both sitting in black swiveling chairs. The window adjacent to them revealed it was nighttime here, as it had been in Spain.

"Our voices?" Tracey mimicked her brother's question, not really knowing what else to say.

The man chuckled again. "I know it can seem kind of intimidating at first, knowing so many people are going to hear your voices, but once you get going, the words will flow naturally."

The Sassafrases still had no idea what he was talking about, and their faces showed it.

"Oh, you know what? Where are my manners? I haven't even introduced myself. Plus, this was all so last minute, I don't even know your names. I'm Lubanzi LeRoux, the producer and host of the show. And the two of you are?"

"We are Blaine and Tracey," Blaine answered. "Blaine and Tracey Sassafras."

"So very nice to meet you, Blaine and Tracey," Lubanzi said as he shuffled forward across the carpeted room for handshakes. He reached toward Blaine first, and as their hands touched, they were both zapped by a small surge of static electricity.

"Whoa," Lubanzi exclaimed, but still smiled. "Shocking."

Blaine nodded, staring at his raised hand.

"Static electricity is an electrical charge that stays in one place," the twins' local expert for this location said. "That's what just happened in our handshake, Blaine. It's also what causes your hair to stand on end, the crackle when you take off a sweater, and what causes lightning. It happens when electrons get rubbed off one surface onto another surface next to it."

Lubanzi LeRoux laughed again. "Oh, here I go again. I do this all the time. Especially on the show. I launch off into dialogue about scientific topics. Please forgive me."

"No need," Tracey smiled. "We love to talk and learn about science."

"Wonderful, just wonderful," LeRoux said. "Well, if it's okay with you, Blaine and Tracey, let's make our way to—"

Lubanzi LeRoux's sentence was suddenly cut off when the lights in the room shut off. The twins sat in the dark for a few long moments, both thinking about saying something to their local expert. Before anything came to mind or mouth, the lights flickered back on. Lubanzi still had a smile on his face, but he had added a wrinkle on his brow.

"These surges in power have been happening on and off throughout the night," he said. "I don't know if it's because of the winter weather we are experiencing or if it's something else."

"Winter weather?" Either one of the twins could've easily asked out loud, but then they remembered what they'd learned about hemispheres. It was summer in the northern hemisphere right now, which meant it was winter here in Cape Town in the southern hemisphere.

"It's not snowing or anything like that, but it is cold and rather windy out there," the South African man continued his thought process. "I wonder if that is causing these electrical surges in the building. You know, electricity is basically just moving electrical charges."

"What do you mean?" This time, the twins did ask their question out loud.

"Electricity, or electrical current, is a form of energy carried by tiny, charged particles," Lubanzi explained. "Electricity can be

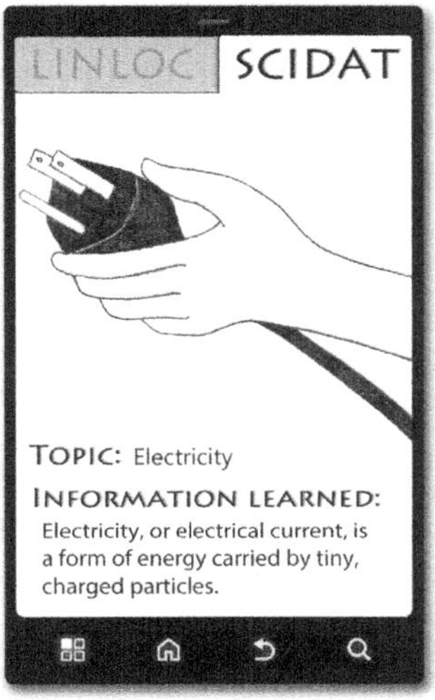

made from different types of energy, such as light from the sun and wind. This type of energy can be turned into light, motion, or sound. The tiny, charged particles can easily flow along cables. So, electricity is used to power our modern-day lives. The flow of electricity is called an electrical current and is measured in amperes, or amps."

The Sassafrases understood what LeRoux was saying, and they were taking precise mental notes so they could enter all this into their SCIDAT apps later.

"Substances that carry electrical current are called conductors," the man continued. "Things like copper wire and aluminum foil are conductors. Substances that electrical current cannot pass through are called insulators. Things like plastic and wood are insulators. Modern-day electrical wire consists of an internal conductor, like copper wire, and an external insulator, like plastic sheathing."

LeRoux took a breath and then said, "All that to say, there are electrical currents running throughout this building, but it is built well and is certainly up to current safety standards. Normally, the outside weather doesn't cause the power to surge like it's been doing. It must be something else."

The tall man shrugged. "Hopefully, it won't happen again, and if it does, hopefully it won't affect the show."

"What show?" The twins asked.

Lubanzi LeRoux looked at the twins like they were joking, but then he realized they were not. "Man, they didn't tell you much of anything, did they?"

The Sassafrases shook their heads.

"Wow, okay, well let me show you!" LeRoux said cheerfully in his deep voice. The tall man led the way out of the room into a hallway. "We're currently on the top floor of the Marble Bridge building in downtown Cape Town. During the daytime, this office building is full of people working in office spaces on all the other floors, but at night, it's just us here on top, running the show."

The twins nodded, still not understanding what show they were supposed to be a part of.

"I am such a fortunate man, and I feel so lucky to be able to do this," LeRoux said genuinely. "I graduated from the University of Cape Town with a degree in physics, and I immediately got a job in research. I liked my job, and I loved physics, but it had always been my dream to be a DJ. So, after years of working, I took a risk and tried my luck in the world of radio. Somehow, someway, with lots of breaks and help from others, I made it!"

"You did?" Tracey asked as the trio continued down the hallway.

The tall man nodded. "And now for about two decades I've been the producer and lead personality for South Africa's twenty-fifth or maybe even twenty-fourth most popular nighttime radio show: Cape Town the Whole Night Through, with your host Lubanzi LeRoux, weeknights from midnight to 6 a.m. on Marble Bridge Radio!"

The twins weren't really sure if they would consider that making it or not, but Lubanzi seemed excited about it, so they congratulated him. At the end of the hallway, a door stood open. LeRoux stepped in first, and as soon as he did, he swung his arms out wide and said, "Welcome to my studio!"

Confounding Circuits

The Sassafrases glanced around the studio. The walls in the room were all covered with spongy, beige-colored acoustic panels. In the center of the room there was a desk topped with two large flat-screen computer monitors, several neat stacks of paper, and a big fancy-looking microphone. To the left of the desk, there was a plush couch already occupied by two rather large white men, sitting silent but smiling. They waved at the twins. The twins waved back.

Next to the couch, there were also two unoccupied stools. In front of each stood a microphone and a music stand.

To the right of the desk, there was a large glass window that looked into another room. In that room sat another man behind a desk dominated by a huge, lit-up audio mixing board. The walls in his room also looked to be covered in acoustic panels. The tiny guy wore headphones and a ball cap with curly strands of ginger hair whipping out. He also waved at the twins. They again returned the greeting.

"That's John Z, my audio mixer," Lubanzi introduced the man behind the glass first.

"And then over there on the couch are Chester and Lenny with 'Chester and Lenny's Good Morning Products," the radio DJ introduced. "And that leads me to why the two of you are here, Blaine and Tracey. We need your voices becau—"

In the middle of the man's sentence, the power cut out again.

"Oh now, there it goes again," Lubanzi said in the dark. They all waited and waited a little more, but this time the power did not click back on. They heard a rustling and LeRoux's phone clicked on, illuminating the room with a soft glow.

"Okay. It's 11:46 p.m. We have fourteen minutes until the show starts." He looked over toward the glass window. "John Z, I'm going to go check the breaker box down in the electrical room. Maybe something is wrong down there."

John Z gave a thumbs up.

"Chester and Lenny, I'm sorry this is happening," LeRoux said, turning to the two big men on the couch. "But don't worry, we'll get everything up and running. You will get the word out about your products."

Chester and Lenny nodded.

"Blaine and Tracey, you two come with me! Let's go down to the basement and check the breaker boxes for the building. Hopefully, we can figure out what's going on, and in the process, we can talk about science. But we gotta go fast! The show starts in fourteen . . . no, thirteen minutes!"

The Sassafrases clicked on the screens of their own phones for a little more light. They followed their local expert out of the studio, down the hall, past the room they had landed in, and to the other end, where there was a pair of elevators.

"How about we take the stairs instead of the elevator?" LeRoux said. He was already pushing a side door open that said EXIT on it. "You know, considering the power's out."

The twins thought that was a good point, so they followed the DJ through the door and down the stairs. Turned out, the building was seventeen stories tall, plus a basement. Going down eighteen flights of stairs wasn't that bad, but the twins wondered what the going up would be like if the power didn't come back on. The basement was a wide, unpartitioned space by what the twins could see with the provided phone light. It was spotted with thick cement pillars and a variety of maintenance equipment. The only room looked small, and it was located in the very center of the open space. A sign above the door read: Electrical Room.

Lubanzi had pulled out a set of keys to unlock the door, but the keys were not needed—the door was standing open. "Hmmm, that's strange," the man mused. "This door is always closed and locked. I wonder why it's open?"

The DJ slowed his walk toward the door just a bit. Was this power outage just an issue of electrical maintenance, or was something more sinister happening here? Lubanzi shone his light into the room and then stepped inside, followed closely by the twelve-year-olds who were now feeling more aware of the shadows than the sparse illumination. The walls of the cramped room were covered by large gray metal electrical boxes interspersed with rows and lines of electrical conduit. Just like the door to the room had been, the door to one of the boxes was also open. Lubanzi LeRoux stepped over to that box and shone his light inside.

"Hmmm," the South African man said again. "I don't think the power surges have had anything to do with the weather outside. I think the problem is here. This is the main breaker box that runs power to the entire building, and it looks like it's been tampered with."

This statement sent tinges of fear to the twins' hearts.

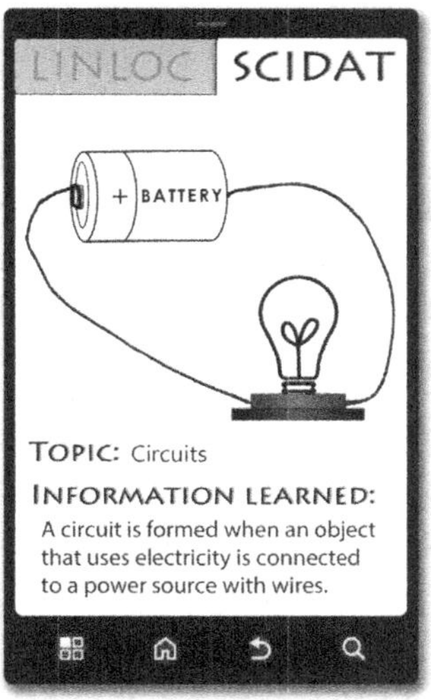

"A breaker box is a critical part of a building's electrical system, as it protects the circuits and distributes electricity to them," he said. "A circuit is formed when an object that uses electricity is connected to a power source with wires. Buildings and homes have circuits to connect the lights and appliances to electricity. When a circuit is complete with no gaps, it allows electricity, or current, to flow around it. When there is a gap, the current or flow of electricity is cut off because part or none of the circuit will receive power."

In the dim light, the Sassafrases could see rows of switches. Lubanzi shone his light up and down the rows and then pinpointed one particular switch with his light.

"See here, the breaker switch that controls the building's main power has either been tripped because of an overload or someone came down here and manually switched it off."

The twins shuddered again.

"A breaker is a safety device that automatically trips off or puts a gap in the flow of electricity to a circuit when there is an electrical overload or overcurrent. This one looks like it was manually turned off, not tripped. I'm not sure who would do this or why. I'm also not sure why there were so many surges before the eventual big shut off, but regardless, let's see if we can't get the building up and running again." The DJ reached his hand up to flip the switch back on.

"Okay, here goes nothing. Let's see if this turns everything back on."

Blaine and Tracey held their breath. The switch was flipped, and then . . . there was light!

"Wonderful!" The local expert exclaimed.

Blaine and Tracey breathed again.

"In a circuit, you can find a few things," the local expert said as he closed the metal door to the breaker box and then led the way out of the room. "You may find a cell, or battery, to supply the power, a switch to turn on and off the flow or power, and components that use power, such as a light bulb, fan, or appliance."

The three walked through the basement, this time toward the elevator. The open space looked pretty much the same in the light as it had in the dimness: spread out cement pillars and a bunch of maintenance equipment. If there was anyone hiding down here, the twins didn't see them.

LeRoux pushed the elevator button. "Items in a circuit can be arranged in two ways: in series or in parallel. In series, the current passes from one item to the next. If a component fails or a switch is turned off, the current stops flowing. In parallel, the current has more than one path. If a component fails or a switch is turned off, the current can continue to flow through another path. In homes, and in our case here in the Marble Bridge building, the electrical current is being carried via parallel circuits. However, if the main breaker is turned off, like it was back there in the box, then the electricity to everything gets shut off."

The radio DJ glanced at the time on his phone again just as the elevator doors opened up.

"Five minutes until showtime," he said as the trio boarded and rode all the way to the top. The Sassafrases were very thankful another power outage didn't happen during the ride.

The three raced down the hall and back into the studio. This time Lubanzi shut the door behind them.

"Okay, Blaine and Tracey, take your places on those two stools over there, and get ready because all of South Africa is about to hear your voices! Well, at least everyone who is still awake and tuning in."

The twins sheepishly walked over and took seats on the two stools behind the two microphones and two stands. What were they supposed to say?

LeRoux sat down at the desk in the center of the studio. He looked at the time again, then at John Z in the audio mixing room, then back over toward Blaine, Tracey, Chester, and Lenny.

"Three minutes 'til showtime," he said with an excited smile. "Let's put on our headphones, and then I'll give some last-minute explanation and instruction."

The DJ put on an enormous pair of headphones himself as the two on the couch and the two on the stools found the provided headphones and did the same. Lubanzi leaned forward and spoke into the fancy microphone on his desk. "OK, friends, if you can hear me loud and clear, say 'Lekker!'"

Blaine and Tracey didn't know what Lekker meant, but they could hear LeRoux's voice loud and clear in their headphones, so they said the word into the microphones on the stands in front of them. Chester and Lenny, who were wearing headphones but who didn't have microphones, simply gave four thumbs up.

"BTW, Lekker is a common word used in South Africa that basically means nice and great!" The tall, happy DJ explained.

"Lekker!" Blaine said again, this time with a little more enthusiasm.

"As I said just a little bit ago," LeRoux now quickly explained, "Cape Town the Whole Night Through is a program that broadcasts live from midnight to 6 a.m. every Monday through Friday. I am the producer and host. John Z is the audio mixer. He's the best in the business, and he makes sure all is well and everything sounds like it's supposed to."

"Lekker!" John Z now said into his microphone.

"Our show consists of five key components," Lubanzi continued. "First, what I'd like to consider witty banter between myself and John Z. Second, we have live phone-in callers. Third, there's the introduction of new music and artists. Fourth, there's dialogue about scientific and popular topics, and fifth, we have products sold by our sponsors live and in the studio."

The DJ now pointed at the two men on the couch. "Chester and Lenny have been sponsors and enthusiastic supporters of the show since we started."

The two couch potato men smiled and nodded in confirmation.

"They are cousins who went into business together years ago, and they are here tonight to sell their new line of Good Morning products. Also tonight, they have committed to give half of the proceeds to a charitable cause. However, at the last minute, both Chester and Lenny came down with horrible cases of laryngitis and they've lost their voices."

Again, Chester and Lenny smiled, nodded, and pointed to their throats as if to confirm what LeRoux had just said was true.

"That's where the two of you come in, Blaine and Tracey," the DJ said, addressing the Sassafrases. "Thank goodness you two showed up tonight. Now we can use your voices to sell Chester and Lenny's new Good Morning products!"

The twelve-year-olds had surprised looks on their faces, which LeRoux read correctly.

"Don't worry. You don't have to have anything memorized," the producer encouraged. "Everything you need to say is written word-for-word on those papers on the stands in front of you."

Each twin looked at the stands facing them and noticed there were indeed ready-to-go scripts waiting for them to read.

"There are two parts that need to be read for each product. There's a section describing the product and a product disclaimer." The local expert now launched into the instruction part of what he needed to say. "I want the two of you to alternate back-and-forth reading those parts. Blaine, you start with the description. Tracey, you start with the disclaimer."

The Sassafras twins gulped and nodded.

"Use a big, clear, energetic voice for the descriptions and a clear but super fast voice for the disclaimers. Does that make sense?"

The twins nodded and gulped.

"After the two of you read for each product, I will follow up with the information about tonight's charitable cause. Does all of this make sense?"

The twelve-year-olds nodded again, this time with resolve. They could do this.

"Lekker!" Lubanzi exclaimed and then consulted his phone once more for the time. "Okay, we go live in thirty seconds! Let's do this!"

It had taken him a couple of days, but he finally had the perfect date set up for himself and Summer. Paul Sims stepped out of the guest bathroom, headed down the stairs, dodging the items stacked there. For some reason, Cecil seemed to use his staircase as a storage space. He went through the living room, out the front door, and across the front porch, being careful not to step on the trapdoor that opened to the slide that led to the basement. He skipped down the steps to the sidewalk on his way to pick up Summer at Mrs. Pascapali's house.

Paul had thought long and hard about how to woo Summer. What kind of date would she appreciate? What kind of time spent together could win over her heart? The go-to stereotypical date was dinner and a movie, but Summer T. Beach was anything but stereotypical. How about dinner and a book? Paul had thought. If he and Summer became an item, then surely Triple S, even with some evidence against him, would look elsewhere for the crimes committed. Wouldn't they? It was at least worth a shot.

It was 5 p.m. right now. He would pick Summer up, and they would walk together to the library. Mrs. Womberfraggle was going to keep the building open a little longer for them today. He had a special little space picked out where he and Summer would each find good books, maybe cozy up a little, and then read, discuss, and exposit the contents of the books. After the library, they would walk to the Right-Handed Pelican Pizza Parlor for dinner. He had hoped to get dinner reservations at a nicer restaurant, but the neighborhood didn't have one. He'd made reservations for 7 p.m., and he'd asked for a back corner table to be set up with candles and a crisp white tablecloth. Surely Summer would love that.

Paul walked confidently down the sidewalk, passing Old Man Grusher's house. The old man was sitting on his front porch next to his miniature poodle. Paul waved and smiled. Old Man Grusher returned the wave and smile.

Paul chuckled to himself, shaking his head. Cecil called Old

Man Grusher's dog the Guardian Beast. "How could he be so afraid of such a little animal? What a loser," Paul thought to himself. He didn't know where Cecil currently was. The weird scientist was usually down in his basement, but when Paul had ventured down there earlier to see if Cecil had any spare toothpaste, his old classmate had not been there.

"Oh well," the museum curator thought to himself as he approached Mrs. Pascapali's house. "Cecil Sassafras's current location is inconsequential. Right now, I just need to amp up the charm. It's time to woo Summer T. Beach off her feet!"

CHAPTER 13: FINDING GOOD MORNINGS

Brilliant Batteries

The show started right on time. "Good evening, friends. You are listening to Cape Town the Whole Night Through with your host, Lubanzi LeRoux." The twins' local expert delivered his introductory line in a deep, smooth voice. John Z played the show's theme music, which lasted about fifteen to twenty seconds. Then the producer and the audio mixer launched into a good-natured and spirited conversation about a range of topics. John Z's voice was a little higher pitched than Lubanzi's, which provided a differentiation between the two.

After a short, almost unnoticeable lull in the conversation, LeRoux said, "And now, a word from our sponsors right here, live in the studio."

He then smiled and pointed at the twins.

Without pause, Blaine began reading the product description in front of him. "Having trouble waking up in the morning? Try new Chester and Lenny's Good Morning products! For a limited time only, Chester and Lenny are offering the exciting Barnyard Buddy! How long has it been since you've seen a sunrise? Do you find yourself so sleepy-eyed in the mornings that those first rays of light go unacknowledged and unnoticed? If you're ready to wake up bright and in the light, the Chester and Lenny's Barnyard Buddy is perfect for you! Place the life-sized, battery-powered, plastic rooster in your room near a window or in a location that receives natural light. Then simply set the alarm, and at the first hint of sunlight the Barnyard Buddy will cock-a-doodle-do you right out of your slumber! Ready to get a booster from the rooster? Try Chester and Lenny's Barnyard Buddy!"

Tracey added the disclaimer, "Barnyard Buddy is not responsible for any other live farm animals that may or may not begin congregating around your house or in your room because of the lifelike call of the plastic rooster."

The Sassafras twins finished reading the script for the Good Morning Product. Blaine had read the advertisement clearly with a perfect balance of charisma and refinement. Tracey had articulated every word of the disclaimer, yet had still delivered it as quick as a whip. They had shocked themselves with how professional everything sounded.

LeRoux's face didn't look surprised, but it did look pleased. It was as if he knew they were going to knock it out of the park. Chester and Lenny also looked extremely pleased, certain that Blaine's and Tracey's voices would sell their products. However, the two salespeople weren't only trying to sell products, they were also here to raise money for a charitable cause, and Lubanzi LeRoux jumped in to share that information over the airwaves.

"Okay, to all our listeners out there tonight who are considering purchasing one of the Good Morning products that are available, take note. Chester and Lenny are pledging to give half of tonight's proceeds to our highlighted charitable cause: Finding Esperanza's Mom."

Wait, what? The Sassafrases reacted without a sound.

The DJ continued. "There is a breaking story coming out of Barcelona, Spain tonight where a young musical prodigy shocked and awed the world with her composition of a breathtakingly beautiful new song. Her name is Esperanza Tambor, and the song is entitled 'Find Me.' She and a pocket orchestra of her young peers performed the song tonight at the esteemed *Palau de la Music Catalana*. It turns out the song also has a message, a plea of sorts. You see, Esperanza is an orphan. She was accidentally abandoned on the streets as a young child. This song is a call to find her mother."

The twins couldn't believe what they were hearing, but it was good news. Lubanzi continued with the announcement. "So listeners and friends of the show, half of the proceeds of every purchase you make tonight of Chester and Lenny's Good Morning products will go to this charitable cause of Finding Esperanza's Mom."

Chester and Lenny were sitting on the couch and smiled like they were so willing and grateful to be part of the cause.

"And now, here on Cape Town the Whole Night Through," John Z cut in from the audio mixing booth, "We introduce to you Esperanza Tabor's song, 'Find Me.'"

The man pushed play and then, in their headphones, the Sassafras twins heard the beautiful song they'd played a few hours earlier being broadcast to the masses. It was wonderful, and they just knew this would help their friend find her mom. Surely there were other stations in other parts of the world that were making this announcement as well.

Esperanza was going to find her mom. Blaine and Tracey

were sure of it.

"He was sure of it. The memory-wiping junior-high class reunion inside the oversized Forget-O-Nator was going to work. The process of making the machine look like a classroom from the inside and a part of the building from the outside had gone splendidly. No one had noticed.

He was here now at his house at 1108 North Pecan Street, where he'd been able to catch up on some much-needed rest. At the same time, he remained alert. Cecil Sassafras had walked down the sidewalk in front of his house not an hour ago, and now here went Paul Sims. Where were his ol' buddies going? They'd both been walking in the same direction.

He walked to a window and saw Paul walking up to that old widow's house. That's where Summer Beach was currently staying. Cecil had not stopped there earlier, though he had given the abode a long glance as he'd passed by. But Paul was stopping . . . and knocking . . . and who was going to answer . . . oh, yes, there was Summer . . . and the two were now leaving Mrs. Pascapali's together.

He scratched his hairless brow. "I wonder where Cecil went. I wonder where these two are going."

"Having trouble waking up in the morning? Try new Chester and Lenny's Good Morning products. For a limited time only, Chester and Lenny are offering the terrific Kick in the Pants Pants! Do you find yourself feeling sluggish in the mornings? Do you find yourself slow-going and in need of motivation? Is what you really need in the mornings just a swift kick in the pants? Then Chester and Lenny's Kick in the Pants Pants are the perfect thing for you! Slip on a pair of comfortable and stylish K.I.T.P.P.s before you hit the hay. Then simply set the alarm and at the appointed time, a battery-powered plastic foot, cleverly built into the back waistband of the pants will pummel, punt, and kick your hind quarters into action! Want to wake up quick? Then put on pants with some real kick! Chester and Lenny's Kick in the Pants Pants!" Tracey read the copy with enthusiasm.

Blaine followed quickly with the disclaimer. "Kick in the Pants Pants are not responsible for any buckets that may or may not be kicked."

Again, they were both having a hard time not laughing as they read their scripts. These ads were hilarious. Were these actual products that Chester and Lenny were selling? They must be, and evidently people would actually buy these sorts of things.

"Remember that half of the money raised tonight on the purchase of these 'Good Morning Products' goes to Finding Esperanza's Mom, tonight's highlighted charitable cause," LeRoux announced. "And a big shout out to those of you who've already made purchases. Thank you for helping Esperanza and enjoy your Chester and Lenny's products!"

"Hey Lubanzi," John Z chimed in. "How about we take a call or two? We have a few waiting on the lines."

"Of course," LeRoux exclaimed. "Who are you going to put through first?"

"How about Dakari the Truckdriver?"

"Yes! Dakari the Truckdriver! One of our most faithful listeners and frequent callers. Put him on, John Z!"

A button or two were pushed over in the audio mixing booth, and then the twins heard an unfamiliar voice in their headphones.

"Top of the evening, Lubanzi and John Z. Thanks for taking my call."

"Of course, Dakari! How goes it out there on the highways and byways?" LeRoux interacted with the caller like he knew him.

"Going great," the man named Dakari said in a gravelly voice. "Just running a load of bricks up the N1 to Jo-burg. But I wanted to call in and tell you fellas, that song you just played by the little orphan girl is the most beautiful song I've ever heard. I thi—"

The call-in truck driver's call was cut off. The power in the building went out. Again.

Were they going out? Was this a date? Is that why Paul had been acting so giddy over the last few days? Did he like her?

Summer was wondering all these things as she and Paul Sims walked side by side down the neighborhood sidewalk. She was pretty sure her old friend had just tried to grab her hand to hold it.

When she'd pulled away a little, he'd just acted like it was accidental, but was it? Was Paul Sims crushing on her?

There had been no need to go back down to the basement to check the breaker box because the lights came back on almost immediately. They'd lost their connection with Dakari the Truckdriver, but as soon as everything was up and running again, LeRoux apologized to the listening audience for the mysterious power surge. He was back on the air with another caller. This time the caller was a woman named Lindiwe the Cat Lady, who had an even more gravelly voice than the truck driver.

". . .and then Freckles shoved Mittens off the countertop while Snowball and Midnight fought over the fish in the dish. All the while, Fluffy was locked outside because Sparkles had somehow locked the door behind herself when she'd come back in! Oh my goodness, it was a real hoot!"

Lubanzi joined Lindiwe in laughter, like he actually thought what she'd said was funny. The twins, however, weren't really sure why the lady had called in, other than to talk about the slapdash exploits of her clowder of cats.

After the call with the cat lady, the DJ again pointed at the Sassafrases, letting them know it was time to read the next product description.

"Having trouble waking up in the morning?" Blaine read. "Try new Chester and Lenny's Good Morning products! For a limited time only, you can get the fabulous Exploding Water Pillow! Do you find yourself in need of liquid encouragement in the mornings? Are the endless cups of coffee and countless cans of energy drinks just not cutting it anymore? Then try Chester and Lenny's Exploding Water Pillow! The days of water beds may be gone, but the age of water pillows has just begun! Our exploding water pillows are plush, comfy, and come in two-gallon, five-gallon, and even fifteen-gallon sizes. Lay your weary head down on the pillow for a wonderful night's sleep. Then to wake the next morning, simply set the alarm, and at the appointed time a small battery-powered charge will occur, causing the pillow to immediately drench you in cold, refreshing water! Start off your day at a brisk pace with a blast of water in your face! Chester and Lenny's Exploding Water Pillow!"

"Exploding Water Pillow is not intended to be a substitute for your daily bathing routine." As Tracey finished the quick disclaimer, both she and her brother again made their best effort to keep from laughing at the commercial. These ads were funny, and the twins were enjoying being a part of the marketing of these creative products.

LeRoux followed the advertisement with the information about Finding Esperanza's Mom, and then John Z had him on the air with another caller.

"And then I was able to successfully grab the ball from the scrum and take off running," Junior the Rugby Player said. "I gained about ten yards and then got my lateral pass off to our fly-half before getting smashed. He broke through and then got it out to our winger, who ran it the rest of the way for a successful try. At nearly the last possible second, we won the match!"

"Splendid, Junior!" LeRoux exclaimed. "Just splendid. Congratulations to you and the boys on your big win."

"Thank you, Lubanzi," Junior answered over the phone. "But that's not the only reason I called in. I also wanted to ask about these Good Morning products. Are they all battery-powered, or is there anything available with a little more juice?"

LeRoux made eye contact with Chester and Lenny as he received this question from the caller. The cousins both nodded, smiled, and gave a big thumbs up.

"Yes, Junior, there are more Good Morning products that are going to be offered later in the show, and I'm guessing different kinds of power sources will be introduced."

Chester and Lenny were still nodding.

"In the meantime, since you mention batteries," the DJ said into his mic. "Let me put on my scientist hat for a minute and briefly chase a rabbit."

Junior didn't answer. The Sassafrases assumed this was because he was already off the line.

"Batteries are stores of chemical energy that can be converted into electrical energy," the local expert talked about the twins'

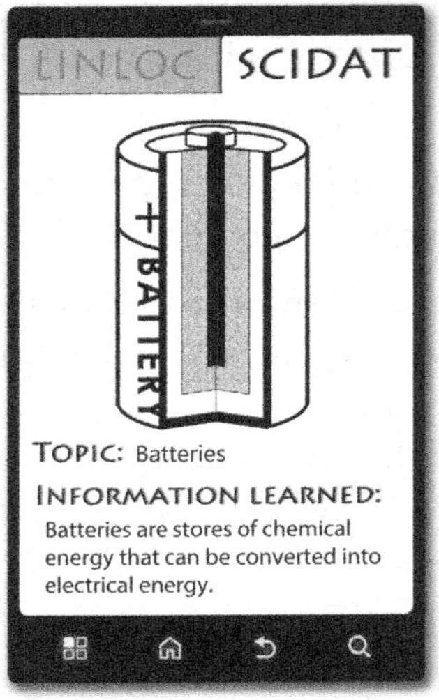

next science topic over the air. "Dry cell batteries are the most common type. These types of batteries contain a paste with electrolytes, which contain charged particles. A chemical reaction takes place, causing the charges to separate and move, which produces an electrical current. Most small batteries, like one-point-five volt batteries, contain a single cell. Larger batteries, like nine volts, contain more than one dry cell. When the electrolytes in a dry cell run out, the battery is spent and can no longer produce an electrical current. Secondary cells are batteries that can be recharged by the electrical current that is generated."

Lubanzi took an inhale and an exhale, and then said, "Ok, now the scientific rabbit has been chased, let's get to the next caller. John Z, do we have anybody waiting on the lines?"

"We sure do," the audio mixer said from his room, but he had a curious look on his face. "It's a first-time caller, and he goes by the name B.B."

"Put him through!" Lubanzi responded emphatically.

The new caller was connected by John Z. Then strangely, the first sound that came from the new individual was heavy breathing.

"Hello? B.B, are you there?" The DJ asked. "You are live and on the air. What's on your mind?"

More heavy breathing from the caller.

LeRoux laughed and then said, "Hey there, B.B, it sounds like you might be sleeping. Maybe you could use one of Chester and Lenny's Good Morning Products!"

"I don't need any good morning products," the caller suddenly snapped. "And it's not just B.B. Those initials stand for something. My name is Bandile Botha. Maybe you remember me? I used to be a DJ myself over on KSAM. I also had the midnight to 6 a.m. slot. I hosted a show with a focus on the multiple uses of cardboard. It was riveting. But then you came along with your fancy variety show and knocked me out of the top twenty-five."

"I did?"

"You did."

"Cardboard?"

"And its multiple uses."

"You were one of the top twenty-five nighttime radio shows in South Africa?"

"I was."

"Really?"

"Are you mocking me?"

"That was not my intention."

Blaine and Tracey looked at each other with faces that said "yikes" as they listened to the awkward back and forth between Lubanzi LeRoux and this caller. Chester and Lenny's faces said the same, as did John Z's face over in the audio mixing room. The twins were pretty sure John Z could cut callers off if he wanted to, but as of right now, he was letting this guy stay on the line.

"Bandile Botha, you say?" LeRoux asked.

"Yes," the man confirmed.

"Yes, I think I remember you. You said you were on KSAM, right?"

This time, instead of confirming, Botha started breathing hard again, which then abruptly broke into a wicked sounding laugh.

"Hey, LeRoux, have you been having any problems with the

power in your building tonight?" The strange man asked through his snicker.

Lubanzi didn't answer. Instead, he just froze, as did all his companions.

"Could it be that someone who doesn't like your show has been down in the basement of your building, messing with the breaker box?"

"Was that you?" Lubanzi asked in alarm.

Instead of answering, B.B. followed with another query. "Could it be that someone is hiding down in the basement of your building right now, ready to mess with the power again or shut it off completely once more if you don't immediately pledge to retire from radio?"

There was another wicked snicker and then the power went out.

Electronic Enigma

He whistled absent-mindedly yet joyfully as he approached the library. He was here to pick up the love of his life. Ernest Womberfraggle had just walked over from the junior high school where he'd spent the day doing some busy work and preparation for the upcoming semester. He was here to meet his wife, Wilma. She closed up the library every day at 5 p.m., but she was known to leave the doors open a bit longer if there were any young, inquisitive minds lingering about. It was a few minutes after five now, and sure enough, the building was still open and lit.

He pulled open the door and walked in. The librarian's desk was immediately inside to the right. That was usually where he would spot his lovely bride, but this evening, she wasn't in her seat. It was a rather comprehensive library for such a small neighborhood, offering rows and rows of books and plenty of nooks and crannies within which to read those books. His beautiful gal was probably roaming these spaces, checking up on anyone who was still here.

Ernest weaved his way around and through rows and shelves of books until he popped out into an open area where there was a semi-circle-shaped stage, suitable for small presentations. It was here that he found Wilma, and she was not alone. Some of his old junior high students were here with her.

"I'm so sorry, Cecil," his wife said sympathetically. "But it's reserved."

"Here and now? Now and then?" Cecil asked.

"Both," Wilma answered softly and then expounded. "It's reserved now for them and later for the others."

His lovely lady stood in the middle of three of his former students—Cecil Sassafras on her right, Paul Sims and Summer Beach on her left, up on the small stage in two comfortable-looking chairs. Ernest Womberfraggle was as confused as Cecil looked, but he didn't feel a need to ask questions. Wilma would sort everything out.

"Now for them?"

"Well yes, Cecil. Paul and Summer have this space reserved now, at least for a few more minutes."

The light and power had come back on rather quickly. When they had, Lubanzi LeRoux pointed at the Sassafrases to go ahead with the next commercial. Though the two were a little rattled because of that last caller and ensuing power outage, the twelve-year-olds proceeded with professionalism.

It was Tracey's turn to read the ad. "Having trouble waking up in the morning? Try new Chester and Lenny's Good Morning products. For a limited time they, they are offering the amazing Magnetic P.J.s! Do you find yourself wiggly and restless when you lay down to sleep? Is this affecting the quality of your sleep so much that you experience the opposite in the mornings? Instead of wiggly and restless, you are laggard and sluggish? Then Chester and Lenny's Magnetic P.J.s are just the thing for you! Made from the finest of fabrics and inlaid with strips of metal, these P.J.s will guard like armor and sleep like royalty! Coupled with the P.J.s, a large powerful magnet will be placed under your bed. At night, the magnet will be set to opposite pole mode, which will invisibly pull at the metal strips in the P.J.'s holding you down snuggly in your bed, prohibiting any wiggles. In the morning, at the time of the appointed alarm, the big magnet will flip to same pole mode, which will fling you up out of bed wide awake with no chance for sluggishness! Try our magnetic push and pull with Chester and Lenny's Magnetic P.J.s!"

"Magnetic P.J.s are not responsible for any metallic household items that may or may not become magnetically affixed to your bed along with you as you sleep." Blaine followed with the quick disclaimer.

Again, even though the mood in the studio was currently a little tense, the Sassafrases couldn't help but grin at the humor of these products.

Once more, Lubanzi followed the advertisement with the information about how half the proceeds of the Chester and Lenny's products were going to the charitable cause of Finding Esperanza's Mom. He then faced their current situation head on.

"To all our listeners out there who just heard the last caller, and to those who may be just tuning in: yes, it's true. We very well may have a mad man in the building threatening to shut down our show by interrupting the power. Regardless, I am here to tell you now that as long as we have electricity and as long as we can maintain connection, Cape Town the Whole Night Through with your host Lubanzi LeRoux will keep airing and will indeed go the entire night through. So, with that said, let's get back to the show."

Lubanzi paused, shuffled the notes on his desk a bit, then leaned back toward his microphone and continued. "Junior the Rugby Player, who was the caller before last, mentioned that he was interested in hearing about some Good Morning Products that ran on something other than batteries—maybe something a little more powerful. Those Magnetic P.J.s we just heard about were battery free, and it sure sounds like there's some power behind the magnetic system they employ!"

As Lubanzi LeRoux spoke, something happened that had not happened since the Sassafrases had been here. Both Chester and Lenny stood up. The big, tall cousins rose from their seats on the couch with looks of determination on their faces. They may both have laryngitis, which was prohibiting them from voicing the commercials for their products, but nothing was prohibiting them from trying to help this radio show they loved and supported.

The two made quick hand motions to communicate they were going down to the basement to see if they could find Bandile Botha and stop him.

The D.J nodded an emphatic thank you toward the salespeople as the two exited the studio. Somehow, someway, Lubanzi continued dialoguing over the air as if all was going perfectly well.

"And speaking of magnets, here are a few scientific facts on the topic. Magnets exert an invisible force on magnetic materials within their field, pulling them towards or pushing them away. Magnets have both a north and south pole, or opposite poles, that attract each other. On the flip side, the same, or like, poles repel each other. Magnets can be found all over the place, like in computer hard drives, speakers, motors, generators, microwaves, refrigerator doors, and evidently also in pajamas!"

LeRoux looked over at his audio mixer. "What do you think, John Z? Are you going to order some of those Chester and Lenny's Magnetic P.J.s?"

"Highly likely," the man behind the glass answered. "And I bet Junior the Rugby Player is going to as well."

"Speaking of Junior, do we have any more callers on the line?" Lubanzi asked. "Preferably callers who don't want to shut the show down?"

"We do," John Z answered. "We have a lot of callers—the Candy Line is waiting on deck to chat with you."

"Ahhh, the Candy Line!" LeRoux exclaimed with a smile. "Yes, they may very well be my favorites. Ladies and gentlemen out there listening tonight, please join me in once again welcoming to the show, the night shift line workers out at the Cape Candies food processing plant: the Candy Line!"

Now the twins could hear a group of people cheering on their headphones.

"Good evening, Candy Line!" Lubanzi said. "I love it that you guys always call in as a big group. What's on your minds tonight?"

A lot of voices could be heard answering, but the loudest voice came through the clearest. "Lubanzi, do you want us to come over to the Marble Bridge building right now? We'll do it! We'll leave the line right now, come over there, and help you out. We'll find that Botha guy and stop him! He won't Botha you anymore!"

Lubanzi laughed and had an honored look on his face. "No no, you guys stay right where you are. Thank you so much for the offer, but I think we can figure this Botha thing out."

"Okay, Lubanzi. But please keep us posted, and if you change your mind, just call out for our help on the radio. We're all listening to your show here at the processing plant. Cape Town the Whole Night Through, baby!"

The loudest voice from the Candy Line then addressed the threatening one. "And if you're listening, Botha, don't you dare shut down the power again! If the show goes off the air even one more time, the Candy Line is going to come for you!"

Lubanzi LeRoux chuckled at the passion and loyalty of his faithful listeners again and then said, "Thank you so much, Candy Line! You all are the best!"

Overlapping the last part of the DJ's statement, John Z maneuvered around on the audio mixing board and started the music to a song.

"Oh, oh! Does everyone hear that?" Lubranzi asked into his mic. "That sounds like a familiar song! One of our favorites here on Cape Town the Whole Night Through. Here is Sveta and the Spark Plugs with 'Crazy Rad Good!'"

She loved pizza—not as much as sandwiches—but pizza was one of her other favorites.

However, she'd lost her appetite. Here they were at the Right-Handed Pelican Pizza Parlor. Paul sat across the table from her, saying something about how much he was enjoying spending time with her. But it was like she wasn't even here.

She wasn't listening to Paul, and she wasn't hungry. All Summer could think about was that look on Cecil's face when they'd run into him at the library. He looked more than sad. He looked more than shocked. What was his face trying to say?

"Having trouble waking up in the morning? Try new Chester and Lenny's Good Morning products. For a limited time only, Chester and Lenny's are offering the spectacular Roll-a-Sheets! Do you find yourself rolling over and going back to sleep in the mornings instead of climbing out of bed? Is it just too hard to pull off those cozy sheets and warm blankets? If so, then Chester and Lenny's Roll-a-Sheets are the perfect thing for you! Roll-a-Sheets are made from Egyptian cotton and come in an assortment of sizes and colors. Additionally, these wonderful sheets come equipped with a large electronic roller that can easily be attached to the end of your bed. Simply set the alarm, and at the appointed time, the roller will quickly and effectively roll the sheets plus any blankets, quilts, or duvet covers up off your body, leaving you feeling cold, vulnerable, and ready to tackle the day! Act fast and order now because just like the sheets, this offer will soon be roll, roll, rolling away! Chester and Lenny's Roll-a-Sheets!" Blaine read enthusiastically.

Tracey dropped her voice and read quickly. "Roll-a-Sheets are not responsible for any injury or harm that may or may not be caused to pillows, pets, or stuffed animals."

The Sassafras twins continued to deliver Chester and Lenny's ads and disclaimers like seasoned professionals. They were having a blast being on this nighttime radio show. They had doubted the show at first, but now they could easily see how it could be one of the top twenty-five most popular radio shows in South Africa from midnight to 6 a.m. Lubanzi was great. John Z was great. And, to top things off, the power hadn't gone out again, at least not since the Candy Line had threatened Bandile Botha.

But where were Chester and Lenny? The two big laryngitis-stricken salespeople had not returned to the studio yet. Had they run into B.B.? Either way, the show continued. John Z had just put another caller through: a younger sounding lady by the nickname of Kati the Go-Getter. She and Lubanzi were conversing right now.

"Sure, all these Good Morning products sound great," she said. "But for me, waking is not a problem at all. Why? Because there is so much to do! So many problems to solve! So many goals to accomplish! So many tasks to complete! No, waking in the morning is not the problem. It is going to sleep that I need help with. Do Chester and Lenny have any products that can help me with that?"

"Wonderful question and good point, Kati," Lubanzi responded. The DJ then looked toward the Sassafrases and made a hand motion instructing them to look through the rest of the papers to see if there was a script for any products that could cause drowsiness.

The twelve-year-olds did this quickly while LeRoux continued to chat with Kati the Go-Getter. The twins leafed through the scripts in front of them and saw more Good Morning products, like, Thumb Jumper Cables, Robot Bed, and Marching Band in a Box. There didn't seem to be any products that could help with . . . oh wait, here was one that was dubbed a Good Night product, something called the Boxing Glove Bed Post.

Tracey gave a quick thumbs up to LeRoux, who responded with a smile and a thumbs up of his own. "Okay, Kati the Go-Getter, I think we've found just the thing for you! Stay tuned for the next Chester and Lenny's product!" Lubanzi said as he got off the call with the high-energy young woman.

He then pointed at the twins again, instructing them to go ahead with the product description they'd found.

Tracey slowed down her voice as she read. "Having trouble falling asleep at night? Try new Chester and Lenny's Good Night products. For a limited time only, Chester and Lenny are offering the incredible Boxing Glove Bed Post! Do you struggle with insomnia? Do you find yourself wide-eyed and alert when it should be time to sleep? Do you long to be knocked out cold instead of awake, antsy, or anxious? Then try Chester and Lenny's Boxing Glove Bed Post! This ten-ounce, spring loaded, electronically powered boxing glove can easily be affixed to all styles, shapes, and sizes of bedposts. Situate yourself in front of the boxing glove. Select the type of punch you'd like to receive, whether jab, hook, cross, or uppercut. Then press the 'Let's Go!' button. In approximately seven seconds, the boxing glove will punch you in the noggin with enough force to knock you out and into at least eight hours of restful sleep. Don't fight, instead go night night! Chester and Lenny's Boxing Glove Bed Post!"

"Boxing Glove Bed Post is not intended or designed for use anywhere other than a bedpost and is not meant to be used as a sparring partner and-or a self-defense tool against friend, foe, or older siblings." Blaine read fast.

"There you have it, Kari the Go-Getter and the rest of our listeners!" Lubanzi smiled. "Chester and Lenny are selling more than just Good Morning Products. And also, don't forget: half of the proceeds raised by the purchase of tonight's products go to Find Esperanza's Mom, which is this evening's charitable cause."

President Lincoln's Ever-so-brief Presentation on Chemistry had gone so well on the stage at the library a few days ago that Cecil had thought maybe it would be a good place to meet his niece and nephew when they zipped back after their study of physics. Evidently, however, the space had been reserved.

But that was not the reservation that had him out of sorts. It was the reservation he had witnessed already in motion with his own eyes and felt the pain of in his own heart. The space had been reserved tonight for two of his old junior high classmates. It had been reserved for Paul and Summer. They had been sitting there in two comfortable-looking chairs, side by side, small-talking about books. Plain and simple, the two had been on a date. He felt betrayed by Paul, who he thought had been trying to help him woo Summer.

He had missed his chance. Cecil had finally realized that he loved Summer. In fact, he had loved her all along, but he'd waited too long to reveal his love to her. Now she was quite possibly falling in love with someone else.

The scientist walked up the sidewalk back toward his house with head down and shoulders slumped.

"Both of them are?" John Z asked.

Lubanzi nodded. "They sure are," the DJ said into his microphone. "Both of those last two Chester and Lenny's products are electronically operated."

"And that has you wanting to talk some science, doesn't it?" The audio mixer said, sensing what his friend was about to move from brain to microphone.

"It sure does," LeRoux confirmed. "Electronics contain a switch, which when turned on, allows electricity to flow through the machine. Electronics then use a collection of electrical components to control the way the electrical current flows around a circuit, allowing certain tasks to be completed. We call these electronic circuits."

"And what are some components of an electronic circuit?" John Z mused.

"I'm glad you asked," Lubanzi said with a smile. "Transistors, capacitors, diodes, and resistors are all components of an electronic circuit. Transistors are electronic switches. Capacitors store electrical energy and release it when needed. Diodes allow electrical current to flow in only one direction. Just a little-known fact about diodes: a light-emitting diode (or LED) glows when current flows through it. And then here are a few things about resistors. To clarify, resistance is the ability of a substance to restrict the flow of electricity. This is a key factor in electronics because when a substance resists electricity, some of the electrical energy is converted into heat or light. So resistors in electronics use a color-coded system to show how much resistance they give. A variable resistor changes the amount of resistance it provides. A thermistor is a heat-sensitive resistor."

"Wow, that's a lot," John Z remarked. "Sounds like it could be confusing."

"Yes, electrical circuits can be confusing," Lubanzi confirmed. "That is why scientists often draw diagrams of a circuit using different symbols for the items in the circuit. This helps to explain everything."

"Interesting," the twins thought as they sat on their stools and listened to their local expert continue to talk and teach.

"So those last two Chester and Lenny's products that were both electronically powered are guaranteed to have comple—"

Suddenly, everything went dark again. LeRoux's voice dropped off the air. The lights clinked off. The power was out. And this time, it didn't flicker back on.

CHAPTER 14: AT THE BOTTOM OF A CAMBODIAN WELL

Engineering a Design

Junior held a rugby ball. Lindiwe corralled all her cats. Dekari u-turned his truck to drive all the way back to be here. Kari ran here, wide awake and determined for a solution to be reached. Chester and Lenny stood still, unable to speak because of laryngitis, but doing fine other than that. John Z, Lubanzi, and the Sassafras twins were here instead of up in the studio voicing the show. The candy line had shown up as promised—an army of people wearing white aprons, plastic gloves, and hairnets waiting. The arrival of the faithful listeners and callers wasn't surprising. After all, Cape Town the Whole Night Through was one of the top twenty-five most popular overnight radio programs in South Africa.

Everybody congregated in the basement of the Marble Bridge

Building and standing in the middle of everyone was Bandile Botha. The twins thought he looked a little like a smaller version of their local expert, except for his demeanor.

When the lights had come back on, he had apologized for what he had done, and Blaine and Tracey believed him. He had admitted to being the one who had been messing with the breaker box, affecting the power and lights in the building. He had confessed after Chester, Lenny, and the other frequent callers had confronted him. The question now was, would all these loyal listeners forgive Botha for sabotaging their show?

In their travels this summer, Blaine and Tracey had run into a fair number of malicious villains. They had also run into just as many courageous heroes. In the basement of the Marble Bridge, the surrounding people had collectively forgiven Bandile Botha. And there is nothing more courageous than forgiveness—at least that's what Blaine and Tracey thought.

The group had piled back into the studio to finish the show. Chester and Lenny sold an ample amount of Good Morning Products and had, therefore, raised an ample amount of money to help in Finding Esperanza's mom. The best news came in at 5:58 a.m., just before the show ended. Esperanza's mom had been found! Any remaining funds would go toward travel and setting the two up in an apartment in Barcelona where Esperanza would continue to play with the symphonia. This news thrilled the Sassafras twins, knowing the mother-daughter reunion would be wonderful.

After the show, the twins had gone back down to the basement to enter all their SCIDAT information from Cape Town into their phones.

Then, they had opened up LINLOC to find: Longitude 13° 23' 15.0" N, latitude 103° 49' 04.2" E. The coordinates would land them somewhere in or near Siem Reap, Cambodia! Here they would study the topics of engineering design, bridges, and types of simple machines, all with the help of a local expert by the name of Jaya Tang.

Through swirls of bright light, the siblings zipped until they reached the precise longitudinal and latitudinal spot with a jerk. Their specially designed three-ringed carabiners automatically unclipped from the lines, and down the two slumped with zero strength or sight.

As feeling returned, the Sassafras twins found they had landed in . . . the bottom of . . . a pit. A deep pit.

They both tilted their heads back to look up at a circle of light way up above them. The almost perfect circle of daylight was probably twenty-five feet up. Maybe they weren't in just any pit. Maybe they were in a well? The ground they were standing on was slightly wet and a little muddy. But if it was a well, wouldn't it be full of water?

Looking up proved to be easier than looking down or around. It was very shadowy, so they really couldn't determine how wide the new space was. Blaine reached his hands out in front of him and stepped forward to see if he could touch a wall, but it was his foot that hit something before his hand.

"Oomph," the twins heard a person exhale.

"Oh, my cow. I'm so sorry," Blaine yelped apologetically.

"Huh . . . what?" a teenage male said in an accented voice. He sounded startled and groggy, as if he had suddenly just been awakened.

"Oh no," the voice continued. "They threw more people down here? I'm sure they're just joking. They have to be."

Blaine and Tracey obviously did not know what the boy was

talking about. Maybe he was half asleep?

"How many of you are there? Oh, I can sort of see you. There are two of you, right?"

Evidently, he could see them in the dimness.

"Yes, there are two of us," Blaine answered for himself and his sister. "We are the Sassafrases. Blaine and Tracey Sassafras. We are twins, but I'm older by five minutes and fourteen seconds."

Tracey rolled her eyes. Why were those five minutes and fourteen seconds so important to her brother? Important enough to include in a greeting with someone they just met and still couldn't see.

"Nice to meet you, Blaine and Tracey Sassafras. My name is Jaya Tang, but a lot of people just call me J.T."

As he completed his part of the introduction, the twins' new local expert stood to his feet, making himself more visible. Tracey thought he might've been about as tall as they were.

"Sorry again for kicking you," Blaine said, still feeling bad about how he'd located Jaya.

"Oh, that's okay, Blaine," the teenager said. "I must've just laid down and dozed off for a second. I didn't even hear the two of you land down here. Did they throw you in or lower you down with the rope?"

"Huh?" the twins responded.

"Let me just apologize for my brothers," J.T. continued. "They can be pretty rude, but I really do think they're just joking. Although I'm not really sure why they threw the two of you down here. With me, it makes a little more sense, but with the two of you, I don't know what they were thinking."

"Your brothers?" Tracey asked.

"Yes, my ten older brothers," Tang said. "I don't think any of them have ever liked me all that much, but I never expected them to throw me down in a well. Surely, they're just joking."

"Wait, your ten older brothers threw you down here? Down at the bottom of a well?" Tracey asked. "Why would they do that?"

"I'm not really sure," J.T. answered like he truly wasn't sure. "Maybe it's my new bicycle?"

"Your new bicycle?" Blaine quipped, wondering what that would have to do with anything.

"Well, see, I've always loved math and science. Growing up, I often talked about wanting to be an engineer or physicist and maybe work for NASA someday. That always seemed to delight my father. So much so that he bought me my very own NASA T-shirt and matching space-style bicycle. Now, I don't know that I would ever really move to the U.S. to work for NASA. I think I'd rather stay right here in Cambodia and use anything I learn to help my home country, but I've always been a big fan. So my father bought me the T-shirt and bike as a sort of affirmation."

"Why would that make your brothers want to throw you in a well?" Blaine asked, still not understanding how any of this connected.

"We are a poor family," J.T. continued. "So for my father to buy any of us something, like our very own brand-new bike, was a big deal. I saw his act as kindness, but my brothers probably saw it as favoritism."

"Ohhh," Blaine now said, sort of connecting the pieces.

"I think I've probably always annoyed my older brothers," the teenager continued. "They are all rice farmers. I'm the only one of my brothers who's ever shown an interest in math and science. They've never understood me, and to be honest, I probably don't understand them. I would come up with scientific solutions for issues I perceived, but they never wanted to hear any of it."

"Like what?" Tracey asked.

"Well, Cambodia often struggles with drought," J.T. said. "As rice farmers, my brothers need water to cultivate their crops. Any level of drought is bad for them. I once read about someone who used old bicycle parts to make a windmill. So last dry season, I was going to use this idea and parts from my bicycle to design a windmill for my brothers to use to water their fields, but they refused to listen to me. They don't understand me when I use scientific terms. It's like I'm speaking a foreign language to them. I think I especially annoy them when I see a problem and begin talking through the engineering design process."

"The engineering design process?" Tracey asked.

"Yes," Tang confirmed. "Engineers use science, math, and their own curiosity to design products and processes that meet needs and solve problems. The way engineers do this is through an eight-step process they call the engineering design process."

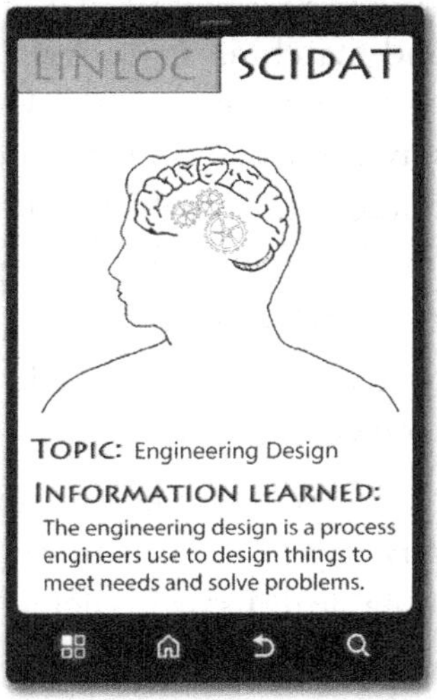

"What are the eight steps?" Blaine asked.

"Wow, you Sassafrases must love science as well!" The young local expert said with some authentic sounding joy in his voice.

"We do!" The brother and sister answered emphatically.

"It's pretty refreshing for me to be around people who share this love of mine."

"Us too," the twins said in unison.

"You really want me to expound on the eight steps of the engineering design process?" The Cambodian asked, just to make sure.

"We definitely do." This time Tracey answered for them both.

"Step one: identify the need or problem," Jaya Tang started. "The process begins with identifying the design goal for a new product or process. The engineer will ask if the design is trying to solve a problem, improve an existing product or process, or meet a need. Step two: do research. Once the engineer has the information from step one, they will do a bit of research. He or she may interview users, collect information about the topic, or investigate how the problem has been addressed in the past. Step three: define the design goal. Once the engineer has completed the research, he or she will revisit the design goal and rewrite it based on the research information. The purpose here is to make sure the design goal can be achieved and that the design will truly meet the need or solve the problem."

"Step four is to brainstorm for ideas. Now that the engineer has a clearly defined goal, it's time to brainstorm for ideas. The engineer will come up with lots of ideas, some practical and some crazy, and will write all of those ideas down. Step five is to select the best idea. After the engineer has brainstormed, he or she will look over the list and evaluate which idea will be most likely to meet the design goal."

"Wait," Blaine interrupted. "So it sounds to me like this is exactly what you did. You identified the need and the problem with the drought and your brothers' rice fields, and then you worked through all these other steps and came up with this bicycle windmill thing."

"That's right," J.T. confirmed.

"Blaine, you interrupted," Tracey slightly scolded. "J.T. wasn't finished yet. There are still three more steps."

"Indeed, there are," Tang confirmed.

"Oh, sorry J.T. My bad," Blaine apologized.

"No problem, Blaine," the local expert said without guile.

"There are three more steps to the engineering design process, but with the windmill, I never ended up working through those."

"What are they?" Tracey asked. "What are the last three steps?"

"Step six in the process is to build a prototype. A prototype is a working model that allows the engineer to test the idea selected. This is often done in collaboration with other engineers so that the most appropriate idea is selected. Step seven: test the prototype. This is where the engineer takes the prototype and tests it to see if it will meet the design goal. Then, last but not least, is step eight, which is to redesign it if needed. Frequently, the prototype will not pass the test, so for the last step, the engineer will go back and redesign the prototype. After necessary changes are made, the engineer will test the prototype again. Steps seven and eight will be repeated until there is a design that meets the need and solves the problem as perfectly as possible."

"So you never got the chance to build the bicycle windmill?" Blaine asked.

"I did not," Jaya Tang confirmed. "When I reached step five and had selected the best idea, I presented the idea to my brothers, but they had no interest in moving to the next step. None of them wanted to help me build a prototype. And actually, now that I think about it, maybe that's why they threw me down here in the well. That, coupled with the fact that I was checking up on them."

"What do you mean?" Tracey asked.

"I was at home earlier today with Father, and he asked me to go out into the fields to find my older brothers to make sure they were actually working. I was happy to do it because I was thinking, not only could I help Father out, but I could also present my windmill idea again."

Blaine and Tracey nodded, even though their nods were probably not seen down here in the shadows at the bottom of the well.

Busting Bridges

"It took me a while, but I found my brothers not working. They were all just sitting around lazily, no doing a thing," Jaya said, "And if they don't work, we might not eat."

"That's not good," Blaine interjected.

"No, it's not," J.T. agreed. "They were surprised and caught off guard by my arrival, and I was frustrated by their apparent laziness. We started arguing. They said they were working, and that I didn't understand the trials of rice framing. I told them it sure didn't look like it and I was going to report them to Father. They called me a tattletale and a spoiled brat. I told them I actually wanted to help them, and I tried to present my windmill idea once more, but they didn't want to listen. They called me a dreamer and a science geek. I got frustrated, told them I was probably going to end up being their boss someday, and then I turned to go back home. That's when they grabbed me and threw me down into this well. Most of them were laughing as they did it, so surely it's all a big joke, right? Surely they're just messing around with me and with the two of you, like big brothers often do, right?"

Blaine and Tracey wanted to have an answer for J.T. but didn't. All at once, the end of a rope came tumbling down into the well and nearly hit Blaine right in the head.

"Hey, dreamer! It's time to climb up out of your hole!" a voice shouted down from the top.

The three looked up and could see the silhouette of a person up at the edge of the circle of light.

"See, I knew they were just joking!" Jaya exclaimed in optimistic relief. "I knew they wouldn't leave the three of us down here!"

Climbing the rope hand over hand and pushing up the wall with their feet, the three mud-covered adolescents emerged from the shadows of the well into the bright light of a warm, sunshiny day.

Even with some mud on his face, Jaya was a handsome Cambodian teenage boy who was about their height. His taller brothers were dressed plainly, some wearing wide-brimmed straw hats and some holding the simple tools of a rice farmer. The Sassafrases counted nine boys, not ten. And those nine stared at the three of them mischievously.

"Where—" J.T. was cut off by one of his brothers.

"It doesn't matter," the brother said, stepping forward. "You blamed us for not working, though you've never worked a day in your life. So now, we're going to put you to work, you and your two little friends here."

"What are you talking about, brother?" J.T. asked in confusion. "I work. Maybe not in the rice fields with you guys, but I work. I'm trying to use math, science, and engineering to—"

"I'm talking about actual work, dreamer!" his brother said, cutting J.T. off.

"What? Are we working with you today in the fields?" J.T. asked.

A snicker was his only answer.

"I'm fine with that," J.T. said, like he actually was. "I think I would really enjoy hanging out with all of you guys and working with you."

"No, you will not be working with us," he said with a sneer. "You're going to be working with them."

The brother pointed to his left. The three who'd been at the bottom of the well looked out across the field in the direction Sotha pointed, seeing a large group of people on the other side. The men and women carried tools of some kind, such as shovels, pickaxes, or wicker basket backpacks. The strongest-looking man of the group currently strode across the field toward the Tang brothers and Sassafras twins. With authority and purpose, he reached them in no time.

"These three?" The big man asked as he pointed toward J.T., Blaine, and Tracey.

The brother nodded.

"They'll work for free?" The man added a question to his question.

He nodded again.

"You're sure of it?"

Now there was a collective nod from all nine of J.T.'s other brothers.

"All right then, let's go," the big man said, looking at Jaya and the twins.

"Wait, what?" J.T. asked the man, confused. "What are you talking about? Why are we going to work with you?" Then he turned. "Brother, why are we going to work with him?"

Instead of answering, his brothers just turned their backs and looked the other way.

"No more questions!" The big man said gruffly. "My name is Kosal. I am your new supervisor. Your brothers said you would work for me for free. So for free you will work! Let's go!"

Jaya Tang seemed to be in shock as he followed the big man across the field, away from his brothers and toward the crew of workers. Blaine and Tracey followed as well—the situation seemed more than a little off to them.

They reached the group of people, and the three of them were immediately given tools. J.T. was given a shovel, Blaine was given a pickaxe, and Tracey was given a wicker basket backpack. One young lady did at least seem to be curious about the three new workers. She appeared especially intrigued by J.T. She stood out in her scarlet-colored clothes while the rest of the group just seemed to be wearing dust-covered attire.

Kosal made a motion for everyone to follow him. With a few

grunts and groans, that's exactly what the group of dusty and dirty workers did. They formed a line from the field into and through another field and then through another field.

The fields were large sectioned-off squares. They walked along well-worn and slightly elevated paths at the edge of the fields. The fields themselves were empty, muddy parcels of land. "Are these rice fields?" the Sassafrases wondered.

After traversing the edges of a couple more fields, the twins now saw some green foliage and a handful of old-looking structures made of stone in the distance. Were they walls? Buildings? Monuments? Or were they some kind of mixture of all these things? Blaine and Tracey didn't yet know, and they couldn't quite tell from their current vantage point. They wanted to ask their local expert who was walking in front of them, but nobody else in the single-file line of workers talked, so they chose not to either.

After another fifteen minutes of walking, the twelve-year-old twins now found themselves in what felt like a completely different setting. It was almost as if they had stepped back in time, or maybe into the pages of an old-world history book. Whereas before they'd been in open fields, now they were in the middle of a jungle. The jungle and the stone structures they had seen before from a distance were not two separate entities. They had become one.

Green moss covered much of the stone like the fuzzy fur of a storybook monster. Enormous trees could be seen growing up out of the structures, their myriad of tangled roots spilling down over the stone like the tentacles of mysterious wooden octopi. Faces etched in the stone peeked out at them from the structures, some probably imagined and some actual faces and figures chiseled in the shape and appearance in the rock itself, but partially hiding behind green leaves and slithering vines.

The twins gasped in wonder and J.T. looked back in response. "Angkor Wat," the local expert said.

"Anchor what?" Blaine responded, not understanding.

"No, no," Tang chuckled.

"Angkor Wat," J.T. stated again. "That's where we are. These ruins are a part of Angkor Wat, the pride and joy of Cambodia. So much so that the Angkor Wat ruins are depicted on our country's flag."

The three continued in their single-file line, on a narrow trail that cut between the jungle-clenched stone structures.

"Angkor Wat is a four-hundred acre complex of sacred land that was part of the ancient Angkor civilization. However, what once was a thriving community and city, for reasons unknown, was lost in the jungle. My guess is that it had something to do with irrigation."

"Irrigation?" Tracey asked.

"Yes," Tang confirmed and then continued. "Most of these old stone structures and buildings were temples, but there was also an intricate system of canals and reservoirs built as a part of the old city. More than likely, it was designed this way because of the weather cycle here in Cambodia."

"The weather cycle?" Blaine asked. The twins knew they were asking their local expert a lot questions, but they couldn't help themselves.

"Yes," J.T. said again in confirmation, like he didn't mind at all. "We have a rainy season and a dry season. In the dry season, there is not enough rain for the farmers. In the rainy season, there is often too much. So my guess is that our ancestors designed this vast Angkor Wat complex, with all its canals and reservoirs, as a solution to that problem. If the reservoirs could hold and store water gained in the rainy season and then use the canals to disperse the water to the fields in the dry season, the problem would be solved, right?"

"Did it work?" Tracey asked.

"I think so," Jaya speculated. "But then again, for reasons

unknown, the city became lost and abandoned. Now, much of the complex is in disrepair. The old rice fields on the outskirts of Angkor Wat, in places like where my brothers work, don't produce as they should. The weather can be fickle. For instance, we are in the rainy season right now, but it's not even raining. That's why I designed that windmill—to help irrigate my brothers' fields. But really, it's not just my brothers that need help. It's all the farmers. It sure would be nice if we could figure out a way to get this old Angkor Wat system back up and ru—"

"Hey! New kids!" Kosul shouted from the front of the line. "No talk. Just walk!"

The brilliant Cambodian teenager sighed but obliged, letting his sentence come to a stop and his feet continue to step. The group proceeded on the narrow trail through the ruins of Angkor Wat. Where were they going? How long were they going to walk? Maybe some of the other workers knew, but the twins surely didn't. At least it was an intriguing trek in one of the more adventurous places they'd visited this summer.

Eventually, the trail transitioned from simply going between all the structures to going in and through them. The Sassafras twins continued to be awed by this wonderfully mysterious place. Yes, the jungle-covered structures were obviously ancient, and yes, some were in disrepair, but actually it was more amazing how well preserved most of them were. As they walked through the entryways and in and through the buildings, each twin noticed how well placed the stones were. There was genius and a purposeful design to all of it.

As the group walked under an especially intricate entryway, J.T. pointed up and said, "This entrance is arched. As a matter of fact, this whole passageway is arched."

The local expert was speaking quieter than he had been before. "An arch uses a curve to redirect the supported weight to the sides of a structure. The wedge-shaped pieces are pushed together by

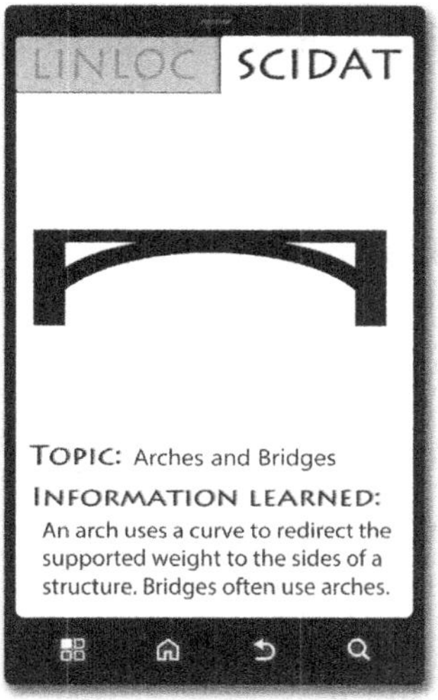

the heavy load, and the structure grows stronger under the pressure because the weight is being transferred to the edges and bottom of the arch."

Tracey, who had both hands free, managed to snap a picture of the arches with her smartphone camera as they walked through the beautifully arched corridor.

"Arches are so strong they can even hold up things like bridges, dams, roofs, or tunnels." Jaya Tang continued as they exited the building, their path now taking them directly onto a stone bridge.

As the Sassafrases looked to either side of the bridge, they could tell it probably used to be situated over water, but now the ground below was pretty dry, except for a small muddy trickle of a stream. They also saw, off to their right, another bridge that was more than likely just like the one they were crossing. It was constructed with stones and employed multiple arches.

"Bridges often use arches to support the deck structure," J.T. said, still talking quietly in a way that only the twins—and maybe the scarlet-clothed lady in front of him—could hear him. "Bridges truly are engineering marvels that carry people, objects, or vehicles across bodies of water or other obstacles as they hold heavy loads and survive extreme weather. They can even be built strong enough to support roads, railroad tracks, and, sometimes, canals."

The twins wondered if there were any canal-supporting bridges in Angkor Wat, or if all the canals here were in-ground. J.T. was right—it sure seemed dry here for the rainy season. The ground could definitely use some moisture, as could the farmers. A working canal system here would do wonders for the agriculture in this place.

Kosal led his single-file line of workers from the bridge back into the jungle. They now traded the cover of stone buildings for the cover of intertwined tree limbs. The pathway under their feet turned from carefully placed stones back to packed dirt. After another minute or two, they reached another bridge, but this one was different. It wasn't made of stone, but rather of wood and rope. It looked old, but not ancient. Whereas the stone bridges and buildings looked like they were built to last forever, this bridge looked like it might not last long enough for them to cross it safely.

Without pause, Kosal stepped foot on the wooden bridge, immediately causing it to bounce. One by one, the workers followed the supervisor without caution. It wasn't a big or long bridge, but it did droop over a fairly deep gully.

"A bridge's span is the distance from one end to the other," Tang said as he stepped onto the trembling bridge. "A bridge's deck is the long flat section that holds the path, or with large bridges, the road, railroad track, canal, etc."

Blaine and Tracey followed their Cambodian friend onto the bridge, joining the others in their bounce.

"This is a suspension bridge," J.T. explained. "Suspension bridges are typically larger bridges that use cables overhead, but in some cases, they can be smaller, like this one."

"Hey, smarty pants," the woman in front of Jaya suddenly spoke up. "You'd better be quiet. If Kosal hears you talking again, he might treat you even more terrible than he treats the rest of us."

The young Cambodian lady said it as a warning, but she also had a smirk on her face. The Sassafrases couldn't tell if she was trying to be kind, unkind, or maybe just sarcastic. Whatever her bent, J.T. paused his explanation. The caravan made it safely across the bouncing bridge and then continued to trudge on.

The ragtag group of workers wound through the jungle a

little while longer. Eventually, they emerged into a more open space where there were yet more old stones. There were two in-ground parallel lines or walls of stacked stones, much of which seemed to be in shambled disrepair. The Sassafrases wondered if this was one of those old canals of Angkor Wat that J.T. had mentioned earlier?

Up in front, the big work supervisor finally stopped and turned to face his sweaty, tired, and weary group of laborers. "Okay, good," both twins thought. "We're finally going to get a break."

"Okay, everyone," Kosal shouted with a sneer. "I hope you're all feeling refreshed and rested, because now it's time to get to work!"

CHAPTER 15: SIMPLE MACHINES SOLVE THE PUZZLE

Simplifying Work

"I think it would be a lot easier if we could use some simple machines. As a matter of fact, I'm sure of it," J.T. said, with sweat pouring from his brow.

"Simple machines?" Tracey asked.

"Machines? Where are we going to get any machines out here?" Blaine added, both of the twins also sweating profusely.

"A simple machine is a tool you can use to help you do work," J.T. explained. "They make the task of lifting or moving an object easier by reducing the forces needed to move the object."

"Oh man, that would definitely make all of this a lot easier," Blaine agreed.

The group of workers were spread out on the two parallel walls of stacked stones embedded in the ground. If a stone block was badly eroded or in disrepair, the crew would chip, dig, or pry it out. They would then replace it with a new stone block.

The Cambodian sun was baking down on them, and the blocks were extremely heavy. Blaine worked with his pickaxe to chip old pieces of stone away. J.T. scooped up the loose stone, dirt, and debris with his shovel. He then put the debris into the wicker basket backpack Tracey wore. Tracey then hiked up and dumped her load into a designated debris pile situated outside the ancient canal. The larger pieces of stone were supposed to be carried by hand to the debris pile.

Once the spot was sufficiently cleaned out, they were supposed to move a new block into that location. The stones were so big and so heavy that some of the adult laborers worked alongside the three for the task.

The young lady wearing scarlet currently worked a pickaxe of her own. She seemed just as interested in the science as the Sassafrases. She listened closely to everything J.T. said.

"You also have complex machines, which are typically a combination of simple machines," Jaya Tang continued his explanation of machines. "Things like drills, jackhammers, trucks, and bulldozers are complex machines. That's probably more what you were thinking when you asked about machines, right, Blaine?"

The Sassafras boy nodded. But he still really didn't know what Jaya was talking about when he was saying simple machines.

Evidently, Tracey didn't either because he asked about it. "So then, what are simple machines?"

"Simple machines are things like pulleys, screws, wedges, levers, gears, inclined planes, and wheels and axels," J.T. explained. "Every complex machine you'll find out there uses a combination of simple machines in one way or another."

"How would any of those simple machines help us out here?" Blaine asked. "I mean, with all of this digging and chipping and hauling and moving we're doing?"

"Yeah, smarty pants," the woman in scarlet chimed in. "How is any of that going to help us with what we are doing?"

J.T. looked like he didn't know what to think about the woman's question. He opened his mouth to answer her, but she jumped in first.

"You know what I think would be more helpful than any of these simple machines you're talking about? I think it would be helpful if you come grab this pickaxe from my hand and give me a break. I know you're smart—I can tell by how you talk—but are you strong? C'mon, Smarty Pants," she said with a smirk.

"I'm not talking about any of us doing the work of another," Tang said, actually taking a step back from the lady instead of toward her. "I'm talking about using simple machines to make this job quicker and easier for all of us."

"How could any of those things make our job easier?" She pouted.

"An inclined plane would make this job of moving these heavy stone blocks much easier," he answered.

"How?" Blaine and Tracey asked in unison.

"An inclined plane, or ramp, is a slanted surface that helps move objects along an incline. Inclined planes can be used to move a heavy object from a lower level to a higher level. If we could build some makeshift ramps here in the canal, it wouldn't be such backbreaking work to move the stone. And we could place the new blocks up in their spots so much easier."

The woman in scarlet didn't seem to be completely following Jaya. Blaine and Tracey were, though, committing all the information to memory in order to enter it into their smartphones at a later time.

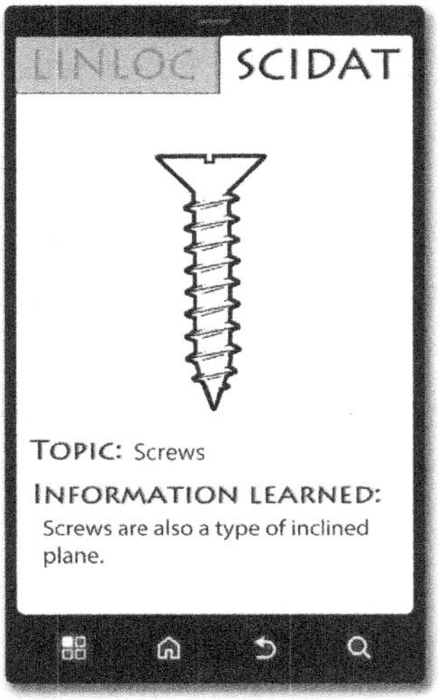

"Screws are also a type of inclined plane," J.T. continued. "It sure would be nice if we had some type of screws or nuts and bolts that would work with stone to help secure all the blocks together better. A screw is an inclined plane wrapped around a cylinder. This plane is also called the thread of the screw. Screws, nuts, and bolts make it easier to drive one object into another object."

"Oh, c'mon, Smarty Pants! What are you talking about? Just come take my pickaxe! Give me a break!"

"I really don't think I should do that." J.T. shook his head.

"If you don't, I'm going to scream for Kosal. I'm going to tell him about all this time you are spending daydreaming and talking."

"I'm sorry, but I'm not going to take your pickaxe. And really, it's not daydreaming as much as brainstorming. I guess I'm kind of working through the engineering design process."

The young lady wearing scarlet then proved she wasn't just full of empty threats. She tilted her head back and with a mix of anger, rejection, and pomposity, she shouted out the big supervisor's name. "KOSAL!"

The boss man was currently supervising another group of workers in the canal and was quite a distance away. However, she had shouted loudly enough for him to hear it. He immediately looked up from the group he was supervising with a look of fury on his face. He then began taking livid steps directly toward them.

"Uh oh," Blaine said under his breath. "This will not be good."

Kosal reached their little group and then stood over them, glaring for a few long uncomfortable seconds. Then he growled, "What is going on over here?"

The scarlet-clad woman pointed at J.T., "Smarty pants here is talking and not working! He even tried to get me to take his shovel and do his work for him!" More fury climbed its way onto Kosal's face.

"And he's daydreaming," the frantic woman continued. "He's coming up with all these ideas on how to make our work easier! Can you believe that?"

"What are you talking about?" The supervisor asked through clenched teeth.

"Simple machines," Tracey suddenly heard herself say, but then she was backed up by her brother.

"Jaya has some fantastic ideas," Blaine now butted in. "You should hear what he has to say about inclined planes."

Kosal now looked like an oversized teapot that was about to screech. "I don't care about simple machines and I don't care about any new ideas," the big man boiled. "Grab your basket, your shovel, your pickaxes, and get back to work!"

"You see, right there," J.T. now addressed the work supervisor. "Pickaxes. A pickaxe is a great example of a simple machine."

"Which one?" Blaine asked.

"A wedge," J.T. answered.

"A wedge?"

"Yes, a wedge. A wedge is a V-shaped object whose sides are two inclined planes. Wedges like knives, zippers, and axes make it easier to separate two objects. For instance, here in the canal, Blaine, you are using your wedge-shaped pickaxe to pull out and separate the old chunks of stone. But if we could design or incorporate some bigger or more resourceful types of wedges here, we cou—"

"Enough!" Kosal interrupted the Tang brother. "If you don't stop talking, I'm going to throw you in a pit!"

"A pit?" J.T. asked.

"Yes! A pit, just like the one I found you in. Just like the one your brothers used."

"But Kosal, I'm just trying to help. Can you imagine how much easier your job would be if all our jobs were easier? Using some simple machines here would benefit everyone, including Angkor Wat and Cambodia. I think it's a wonderful thing we're doing here. We are helping to repair a part of our beloved ancient ruins. But wouldn't it be beneficial to everyone if we did things more efficiently? Even just a little?"

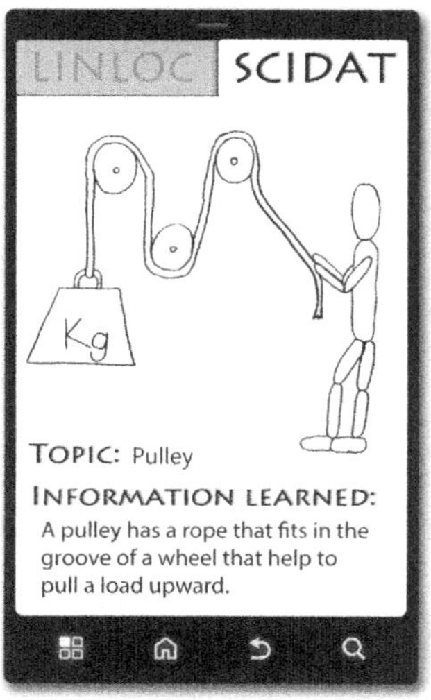

"For instance, what if we set up a rope and pulley system down here in the canal? A pulley is another example of a simple machine. It has a rope that fits in the groove of a wheel. Pulleys make it easier to pull a heavy load directly upward. Wouldn't that work great for what we are doing today? Wouldn't that . . ."

"Enough!" Kosal shouted. "To the pit you go! All three of you!"

It was dark down here. Dark and lifeless. It shouldn't be this way. This place should be full of energy and light and ideas. It should be a whirlwind of inventions, experiments, and scientific fun. Instead, it was silent, still, and dark. The tracking screen wasn't even on.

President Lincoln let out a high-pitched, frustrated, prairie dog bark. This wasn't right. Cecil slept upstairs in his bedroom. Cecil never slept in his bedroom. The wacky scientist was usually down here in the basement around the clock. If he slept, it was more than likely a catnap in between projects, but right now the man was upstairs snoring in his bed.

President Lincoln didn't like this one bit. He needed to do something about it. Cecil Sassafras was his favorite human being. The red-headed scientist was a wonderful boss, a brilliant researcher, and up to this point, had always been hopeful and optimistic. He had also always been joyfully naïve and somewhat socially unaware, which had never caused any major problems until now.

Cecil had finally figured out that Summer Beach loved him. The female scientist had been sending out vibes and cues that Cecil had missed for a long time. But thank goodness, his boss had finally figured it out and had found that he, in fact, reciprocated those feelings of love for Summer. However, worse than missing or misreading Summer's cues, Cecil had more recently misread Paul Sim's cues.

President Lincoln barked out loud again in disgust just thinking about it. He couldn't stand that smug museum curator. He'd swept in here unannounced, acting like he was a genuine friend and acting like he was just here to help. Cecil innocently believed his old junior high classmate. Paul had misused that trust and then tried to steal Cecil's girl. At this very moment, Paul Sims was upstairs sleeping soundly in a bed that was graciously being provided for him by the very friend he was betraying.

Another bark escaped from the perturbed prairie dog. What could he do to help his boss? Was there even anything that could be done?

He crawled over Cecil's desk and tapped the computer's keyboard, which immediately brought the big tracking screen to life, splashing light into the dark basement. Two green dots on the screen were clearly visible. Lights that represented Blaine and Tracey.

President Lincoln smiled amid his current frustration. He was so proud of those twins. It looked like they were currently in the country of Cambodia. If he wasn't mistaken, this was their second-to-last learning location of the entire summer. They would zip back here soon with their complete summer of face-to-face learning finished. This made the current situation even more urgent. He had to get Cecil back to being Cecil.

Yes, she would take the lines and go share the science. She nodded even now, knowing this had been the plan all along. It didn't feel completely right leaving the old neighbor without telling Cecil how she felt. However, she'd be back soon.

And when she got back, she would tell him. She would confess her undying love for him.

Gear Shift

It was dark here, down in the bottom of this pit. This was the second pit they'd found themselves in here in Cambodia. Like their first pit, this one was also a bit muddy, so their landing had at least had some cushion.

"I should've kept my mouth shut," J.T. said sadly. "Why am I always trying to solve everyone's problems? Why can't I just fall in line?"

"No, no, it's okay," Tracey said compassionately. "You were being a good brother and a good friend by trying to come up with solutions."

"Yeah," Blaine added. "Your ideas were all good ideas. The ones you had to make the labor easier in the canal could've helped everybody out! And your bicycle windmill idea was sick!"

"Sick?" J.T. asked, not really understanding the term.

"When Blaine says sick, he means cool or great," Tracey explained and then added, "And I agree. All your ideas have been wonderful, and I bet the windmill really would've helped to irrigate the fields for your brothers."

"Maybe," J.T. said like he believed it could be true, but also in a way that revealed he might've given up.

It seemed even darker down in this pit than in the first one. The three really couldn't see each other at all as they conversed. Kosal had thrown them down into an opening on the floor of a stone structure near the end of the canal they had been working on.

It was an ancient-looking, intricately designed, and moss-covered structure. The hole in the floor they had been forced into looked to be a part of the structure's design. Almost as if it was purposeful, not accidental. However, it didn't really matter what the pit looked like on the outside. They were currently at the bottom, and there didn't seem to be an easy way up or out in the dark.

A bright idea suddenly hit Blaine. "Our smartphones, Tracey! We can use the light from our smartphones!"

"You guys have smartphones?" J.T. asked.

"We do," Tracey confirmed. "Sorry, we didn't think about using them for light earlier."

"No problem," the teenage local expert said. "But let's definitely use them now."

Both Sassafras twins got their handheld devices out and clicked the lights on.

"Wow!" Blaine immediately exclaimed. "I thought maybe we were down in the bottom of another well, but this is no simple well."

"It sure isn't," J.T. agreed.

"Wow!" Tracey added the same exclamation as her brother, but in more of an amazed whisper.

The three found themselves in a large underground room. A seemingly perfectly square room with sophisticated designs and carvings on all four stone walls. Like the hole they had been pushed into, this room felt like it was designed to be a purposeful part of the structure up above.

The twins continued to sweep the light from their phones around the room to further explore this new setting of theirs and to see if there was any obvious way out.

"Wow," Tracey said again in awe. "Look at all the carvings. They are amazing. And they cover pretty much every inch of every wall."

"Hey, look!" Blaine exclaimed. "There is an old bamboo ladder over there on the ground in the corner! Maybe we can use it to climb up out of here!"

The Sassafras boy rushed over and grabbed the ladder, pulling it up out of the mud. It was then that any elation dissipated. Not only was the ladder in some disrepair, it was also clearly too short.

"Well, stink," Blaine huffed. "Even if it was longer, we couldn't lean it against any of the walls to get out because the opening we fell through is at the center of the ceiling. Our only way out of here looks virtually unreachable!"

Tracey thought her brother was right. The room was pretty big. It was probably twenty-five feet across from all four walls and twenty-five feet from the floor up to the small opening. She didn't see any way the bamboo ladder could help them climb out. However, J.T. had yet to share any of his thoughts on the matter.

The Cambodian teenager continued to study the room in silence. The ceiling was a flat stone with no carvings. The floor was flat and muddy. The four walls, however, seemed to have every inch covered with intricate carvings, as Tracey had mentioned. Most of the stone images were of people doing all kinds of different things.

The twins wondered if the carvings all told a story. As the twelve-year-olds continued to look around, they noticed there were more than just images. Hidden among all the people, there were also what looked like stone gears of different sizes scattered throughout the carvings. And was that some kind of writing on one of the walls?

"I've explored pretty much the entire Angkor Wat complex," J.T. said, in a similar tone of awe to what Tracey had been using. "But I've seen nothing like this. Most of the carvings around Angkor Wat depict stories of religion or mythology. There are many that tell stories of battles or give descriptions of the daily life of the ancient Khmer people. But the carvings down here are different."

The local expert paused, took another three hundred sixty degree look around the room, and then continued. "These carvings seem to tell the story of the building of Angkor Wat itself—the story of Angkor Wat and its system of buildings, canals, and fields.

But then there are also all these cogs or gears worked into the carvings. Did you notice that?"

Both twins nodded.

"I've seen nothing like that before anywhere else." J.T. approached each of the four walls one by one. "This wall has carvings and cogs but nothing else."

He then walked directly across the room to the wall opposite the one he'd just described. "This wall has carvings, cogs, and it also looks to be leaking."

"Leaking?" Blaine asked.

"Yes, Blaine, look," J.T. confirmed. "Beads of water are leaking through small holes in the stone. I bet that's why the ground is so muddy."

Jaya then stepped toward the wall that was on his right. "This wall has carvings, cogs, and also a lot of short pegs or shafts protruding from it, plus one longer protruding stone piece that looks like a lever of some kind."

Then he went straight across from that to the fourth and final wall. "And this wall has Khmer script written on it in the middle of all the carvings and cogs."

"What does it say?" asked Tracey.

"Well, it's written in old Khmer," Tang said with a half-squint of his eyes. "But I think I can probably read it."

The teenager squinted more and slowly read the script in the old version of his native tongue. The Sassafras twins really wanted to ask what it all meant as J.T. continued mumbling in a language they didn't understand. He finished reading the script with a puzzled look on his face.

"What does it mean?" The twins asked at the same time.

All at once, the teenager's face went from confusion to clarity.

"Wow, okay, amazing," J.T. started. "Loosely translated, it says, 'Behold the ancient cold-water spring of the Khmer!'"

"Cold-water spring?" Blaine asked enthusiastically.

"Yes," Tang confirmed. "But there's more."

Tracey elbowed her brother. "Let him finish."

"The first part was kind of like an introduction, and the second part sounds like instructions," J.T. explained and then continued with the translation. "Solve the puzzle to prove you're clever. Place the gears and pull the lever. Inside the walls, the wheels will turn. Bringing the substance for which all yearn."

"Wow, okay, amazing," the Sassafrases repeated exactly what J.T. had said a second ago.

"So all this is some kind of puzzle?" Tracey asked.

"Sounds like it," J.T. confirmed.

"We've seen all the gears spread throughout the carvings. Aren't they already placed?" Tracey shared the question rolling in her mind out loud.

"There are wheels inside the walls?" Now it was Blaine. "And what is the substance everyone is yearning for?"

"Water!" J.T. answered. "It all makes sense!"

"It does?" The twins asked, talking at the same time again.

J.T. nodded and expounded. "The script starts by telling us there is an ancient spring. The carvings on the walls depict the building of Angkor Wat with all its buildings, canals, and fields. The complex must've been designed to be fed by this spring. There is currently enough water in parts of Angkor Wat, but not all of it. Like, for instance, the fields of my brothers. If we can figure out this puzzle, then that must mean we can somehow provide water for all of Angkor Wat!"

The twins were glad J.T. understood what needed to be done,

but they were both still very confused.

"I'm going to be honest, J. T.," Blaine said. "I still really do not know what's going on. We got thrown into this pit, and that's not good, right? I mean, sure, all the carvings and the riddle and stuff are cool, but shouldn't we still be trying to figure out how to get out of here?"

"We place the gears and pull the lever," J.T. answered by repeating the particular line from the script.

"How do we place the gears?" Tracey asked. "I mean, they're all made of stone. How can they be placed anywhere other than where they already are?"

Instead of answering with words, J.T. reached out and placed his hand on one of the stone gears on the wall in front of him. He moved his fingers around its edges for a moment and then got a firm grasp of the whole gear. He gave a powerful outward jerk on the gear. With a quick scraping sound, the gear cleanly separated from the wall.

Blaine's and Tracey's jaws dropped. The gears could be pulled off the wall! But that wasn't all. J.T. took the donut-sized gear and walked across to the opposite wall, where he immediately affixed the circular stone to one of the short protruding pegs, the smaller circle at the center of the gear fitting perfectly on the circular stone peg. The twelve-year-old's jaws dropped even more.

"So, my guess is we find all the gears and pull them off. Then we'll attach them to these pegs. The pegs are spread out at different spots all over this wall. The gears look like they're all different sizes. The puzzle is figuring out the precise way all the gears fit on all the pegs, with the teeth of the gears interlocking perfectly with each other. I bet once we get that part right, we can pull the lever which will turn wheels inside the wall. Then, hopefully, somehow the spring water will be released," J.T. said, full of joy with no trace of sadness.

The Sassafrases had pulled their jaws back up, and they were nodding. They didn't know if everything was going to work out just like J.T. was saying, but they were at least understanding his logic.

"So maybe we got thrown in this pit for a reason," Blaine said thoughtfully. "Maybe instead of trying to escape this place, we should try to solve this puzzle. Maybe we really can help everybody get water."

Tracey nodded, agreeing with her brother.

J.T. smiled wide. He looked at the twins with bright eyes and asked, "So what do you say, Blaine and Tracey? You want to figure out this puzzle with me?"

The twins nodded again.

"So just to clarify, a cog is a gear, and all gears are toothed wheels that interlock in pairs, each one helping to turn the next. Gears are often used to consistently drive an object, like a watch, clock, car engine, or hopefully, in our case, an ancient puzzle."

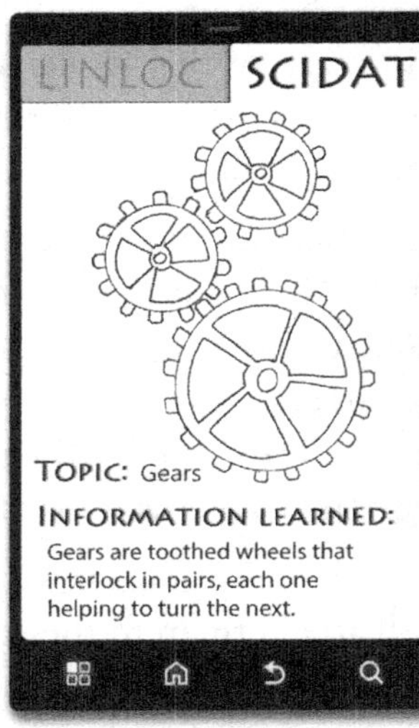

J.T. stood atop the bamboo ladder, giving deeper explanations of scientific topics as he moved around a few of the gears at the top of the puzzle, using the provided light coming from the twins' smartphones propped up on their backpacks. The old ladder was tall enough to reach the highest protruding pegs. It was a pretty rickety old thing, so Blaine and Tracey were both holding the bottom to help keep it as secure as possible. They could tell it was going to be sturdy enough to finish the job.

The group of three had finished a couple iterations of the

puzzle, but in those iterations, there had been a gear or two off because when they had pulled the lever, the wall of gears wouldn't turn. Every gear must have to be in the perfect and intended spot for this puzzle to circulate and work. Either that, or this wasn't even how the puzzle worked.

J.T. placed the last gear, gave a resolute nod, and stepped down off the ladder. All three kids took a few steps back, looking at the entire wall. The different sized gears interlocked with each other and affixed to the stone pegs covered a sizeable portion of the wall. There were 256 gears in all. The big stone puzzle looked cool up there on the wall, but had they gotten it right this time?

"Should we try the lever again?" Tracey asked, almost reluctantly.

J.T. nodded his head and walked over to the protruding stone piece. "A lever, like a gear, is another example of a simple machine, the local expert said. "It is a rigid bar that is free to move around at a fixed point. Levers are frequently used to lift or release things. For example, using a flat piece of metal to pry open a paint can, using a shovel to lift a large rock, or even using a wheelbarrow to move dirt."

"Hopefully, in our case, this lever will act as a fulcrum release, meaning when I pull it, it will release, lift, or disengage from whatever kind of weights or mechanisms are behind this wall and will start our gears to churning."

The twins hoped this would be the case, too. Jaya Tang grabbed hold of the lever and gave it a hearty pull. This time, to the amazement of all three, every gear on the puzzle wall creaked, scraped, screeched, and then started turning.

"Aha!" The Cambodian teenager exclaimed. "It actually worked!"

The Sassafras twins both literally jumped for joy.

"I don't know exactly what's happening there behind the wall, but my guess is the pegs are actually the protruding ends of interior axles. The wheels in the wall must be turning on the axles and our gears are now turning out here on the pegs. You know, wheels and axles are another example of simple machines," J.T. said, his love for science and his current elation colliding.

"The wheel and the axle have two disks, or cylinders, each with a different radius. The wheel and axle are used to rotate things like the steering wheel of a vehicle, a screwdriver to turn a screw, or the interior mechanisms of a giant stone puzzle!"

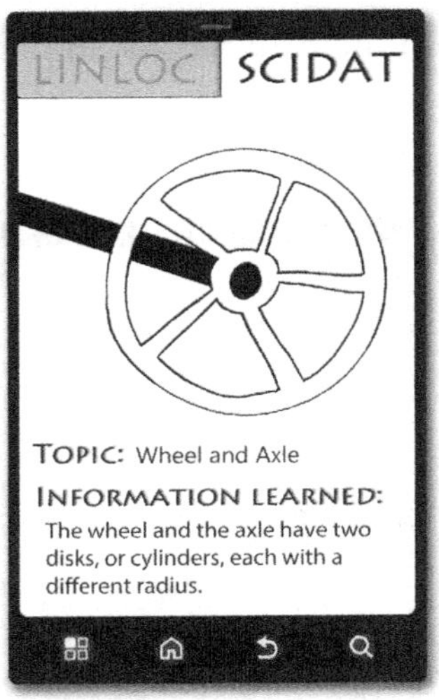

"Solve the puzzle. Prove you're clever!" Tracey happily exclaimed. "Place the gears and pull the lever!"

Then Blaine added, "Inside the walls, the wheels will turn. Bringing the substance for which all yearn!"

Blaine's smile froze in place for a long second as he thought about the last line he'd quoted. "Bringing the substance?" he said. "The substance is water, right? Because this is a spring. But where is the . . ."

The others saw the same thing Blaine was seeing. Those small holes in the wall to their left that had before been only beads of water sneaking out, were now larger holes with streams of water pouring out. As the gears continued to churn, the holes got bigger.

"Behold the ancient cold-water spring of the Khmer!" Blaine announced.

"This is a good thing, right?" Tracey asked with a smile tinged with alarm. "We solved the puzzle. But that's a lot of water, and it's coming faster and faster. Are we going to be okay in this room? It's filling up."

The three looked at each other. Tracey had brought up a good point.

The twins rushed over and grabbed their backpacks and phones before they got lost underwater. J.T. studied all the gears to make sure they were still spinning. The water continued to flow out of the holes faster and faster as the holes got bigger and bigger. The room quickly filled with water. Knee deep, waist deep, shoulder deep, now they were treading in it. J.T. had the wherewithal to grab the bamboo ladder. It was floating, and the three were all clinging to it.

The gears still churned. Volumes of cold water plunged into the room and they were being raised toward the ceiling.

Then something happened that none of the three had been expecting—the fourth wall opened up. Not the wall with the gears and not the wall with the jets of water pumping out, but the wall across from the water-pumping wall.

It split downward and outward from the top. Bright Cambodian sunlight streamed into the room. The three hung onto the bamboo ladder for dear life as they rode on the water out of the room. Where exactly were they going? They were about to find out—at breakneck speed.

"Well, sir, it looks like we are going to have to visit that old neighborhood again."

"Yes it does, DeBlose, yes it does," Captain Marolf gave one stroke to his broom-like white mustache and two to his flattop hair. "Thank you for coming back from your break with such focus. We appreciate the investigative research you did regarding the incident on the moon."

Evan nodded, enjoying this rare praise from his superior.

"Yes, the Rotary Club are criminals who broke the law, but we already have them in custody again," Marolf continued. "It is very clear, per your research, Paul Sims was the mastermind behind the sinister plan. We've got to get him."

"How many agents should I take, sir?" Evan DeBlose asked.

"Just you," the Captain responded. "We'll take a more sensible approach. I think this museum curator is going to be much easier to nab than Bogdanovich and Archer."

"Just me, sir? Can I at least take Q-Tip with me?"

The big barrel-chested captain paused upon hearing the request from his agent. Then he answered, "Okay, take Q-Tip with you. But that's it. Go get him, Agent DeBlose. Go nab this Paul Sims fellow. And then let's be done with this whole thing. Let's close the books on this case."

"Yes sir," the handsome, sandy blonde-haired agent said with a confident smile.

CHAPTER 16: THE MEXICAN FRUIT CART

Modern Machines

Behold, the ancient cold-water spring of the Khmer, indeed! The power of the water was more tremendous than the twins could've imagined. What started as small streams coming out of the wall became an unstoppable current of water flowing forward.

Luckily, the three managed to stay on top of the water as it surged forward—thanks to the trusty floating bamboo ladder. J.T., Blaine, and Tracey rode the flow out of the underground room, through the wall that had opened up, and in the canal they had been working on earlier. Kosal, the lady in scarlet, and the rest of the work crew climbed out of the canal before the wall of water reached them. They looked curiously at the three kids as they briskly floated by. Blaine actually raised a hand and waved to the group of shocked and confused workers, while J.T. and Tracey simply smiled.

They swished and turned through the canals, gullies, and waterways of Angkor Wat. They floated by many of the ancient stone buildings, under several bridges, and then out toward the fields. Thoroughly drenched but kind of enjoying themselves. Eventually, the three friends felt the flow finally slow.

Their feet hit the ground, and they let go of the ladder. The three came to a stop in a certain field. And who was there? None other than the Tang brothers in a line, standing on the elevated pathway with wide eyes and open mouths.

J.T. and Tracey smiled at them. Blaine gave another wave. One brother looked like he wanted to say something, but the words wouldn't come out. All the brothers looked grateful, relieved, and penitent.

"Behold, the ancient cold-water spring of the Khmer," Jaya Tang said to his brothers, gesturing to water. "It looks like all your irrigation problems are a thing of the past—yours and everyone else's! Growing rice and other crops year-round should no longer be a problem. We found a spring designed to water the fields. It was plugged up, but we figured out how to open it."

As Jaya spoke, his voice had no trace of vengeance. "You guys sent us away with Kosal and that work crew, but what you may have meant to be mean or to play a joke has turned out to be a blessing and benefit to us all!"

At this, the brothers fell to their knees overwhelmed with thankfulness, both for the water and for the forgiveness.

He was on his way! On an airplane flying back to the old neighborhood where he'd been a foreign exchange student all those years ago. He was really looking forward to seeing some of his old friends and seeing those twins again, too!

They were on their way. He and his balloon-shaped headed secret service partner. They were in a heliquickter on their way to the old neighborhood, the same neighborhood they'd apprehended Adrianna and Yuroslav in. That apprehension had gone much easier than expected. He hoped nabbing this Paul fellow would be the same.

Longitude 20° 41' 35.1" N, latitude 103° 21' 37.1" W. That's where the Sassafras twins were zipping to next. It was to be a landing location in Guadalajara, Mexico. It was where they would study modern machines, computers, coding, and robots with the help of a local expert by the name of Verano Playa.

"Verano Playa?" Tracey murmured.

"Do you think it's Summer?" Blaine asked as they put on their helmets and harnesses. "We haven't had her for a local expert yet."

"I hope so. I've missed her," Tracey said as they calibrated their carabiners to the coordinates.

With smiles on their faces, the twins clipped their carabiners into the invisible zip lines and they were off through swirls of light.

The smiles remained as the invisible journey ended. When the twins' senses normalized, they found themselves in a small alley piled together on top of. . .oranges? Mangoes? Limes? Watermelon? A combination of all four? It had been evening when they had left Cambodia, but it was daytime here in Mexico.

"Blaine," Tracey started to say. "I think we landed on top of a—"

"*Ay! Mijo! Mija! ¿Qué estás haciendo? Bájate de mi carrito de frutas!*"

A fairly short, fairly round, fairly exasperated Mexican woman interrupted Tracey, and she was storming up the alley toward the twins. She looked to be in her late forties or early fifties. She was wearing an apron spotted with flour and was brandishing a rolling pin.

"Fruit cart," Tracey finished her thought.

"Huh?" Blaine squeaked.

"I think we landed on top of a fruit cart," Tracey repeated in angst. "Her fruit cart! We've gotta get off this thing and apologize."

As the short, flustered woman approached, the twelve-year-olds slid down off the cart and its pile of assorted fruit. Tracey only disturbed one mango. Blaine knocked over a mango, an orange, two limes, and four watermelons.

"*Ay! No!*" the woman said with hands and rolling pin raised as she reached the twins.

"We are so sorry!" Tracey apologized. "We didn't mean to upset your fruit cart.

Blaine nodded and added, "Can we do anything to help? Should we pick the fruit up off the ground?"

"*Ay! Mijo! Mija! ¿Qué estás haciendo? No deberías sentarte sobre mis frutas, deberías las comprar.*"

"I'm so sorry, we don't understand," Tracey apologized.

"She said you should not sit on her fruit, you should buy it," a familiar female voice behind them said.

"Summer?" both twins spun on their heels to find their favorite local expert.

"Actually, today, I'm Verano Playa," Summer announced as she pulled the twins into a happy jumping dance hug.

Summer turned to the little woman with the rolling pin, looking at the three of them.

"*Discúlpenos,*" Summer said. "*¿Cómo podemos ayudarte?*"

"*Pueden ayudarme a servir el desayuno,*" the woman said.

"*Por supuesto,*" Summer answered before turning to the twins and clapping her hands.

The Sassafrases wanted to know what the woman was saying, but they just couldn't understand, and the expressions on their faces revealed it. They looked to Summer for an explanation when an unfamiliar voice suddenly said, "She's asking you to come help her serve breakfast."

The twins spun around on their heels again to see another individual approaching from the opposite end of the alley. It was a handsome young Latino man in his mid-twenties, and he was rolling up in a wheelchair—a really cool-looking wheelchair with a sleek design and a lot of neat hardware.

"Breakfast?" Blaine asked.

The man nodded.

"Matteo!" The woman exclaimed.

The young man in the wheelchair addressed the woman like he knew her very well. "*Mamá, estoy seguro de que fue un accidente. No los obligues a ayudarte.*"

"We'd be happy to help with anything we can," Tracey said. "We are so sorry for damaging the fruit."

"It's really no big deal at all," the man said kindly. "But thanks for being willing to help."

"*¡Vamos!*" The short woman shouted, grabbing a few pieces of fruit off the cart and imploring the others to do the same. She then turned and exited the alley from the same direction she'd entered.

"Matteo! It's so good to see you!!" Summer gave the young man a fist bump, complete with explosion fingers.

"You too, Summer, or should I say Verano," Matteo lifted his eyebrows. "Where is El Asistente?

"Oh, Ulysses is going to meet us at the arena," Summer answered.

The twin's heads swiveled back and forth as they tried to follow the conversation. Matteo somehow knew Summer? Ulysses S. Grant, or rather El Asistente, was in Mexico? And what arena were they going to?

"But right now, we'd better go and help your mother," Summer said with a smile.

The man in the wheelchair nodded. The four of them followed the fruit-carrying, female scientist on her way out of the alley. The Sassafrases also carried their fruit in their arms. The young man piled his fruit on a tray he had on his electric wheelchair.

Summer added a little skip to her walk and said, "Matteo, I'm so sorry. These are my dear friends Blaine and Tracey Sassafras. They are here in Guadalajara to learn more about physics. I was hoping we could take them with us."

"Of course," he nodded, turning to the twins. "My name is Matteo Perez, and that bossy, yet delightful woman is my mother, Solana. She runs several successful food carts here in Guadalajara, and I help her. Also, I happen to love physics."

"Super cool," Blaine said authentically.

The small group exited the alley and entered a wide plaza full of people. Most of them were sitting at makeshift tables, eating. There were businesspeople, construction workers, and students all sitting, scarfing down plates of scrumptious foods and cups of colorful drinks. The morning sun rose over it all.

Summer "oohed," and "aahed," as she said, "It smells delicious!"

Blaine and Tracey nodded as they saw a grouping of food carts at the center of the plaza. Solana Perez was stationed there, beckoning frantically for the three to join her. Blaine, Tracey, and Summer followed Matteo as he knifed through the long lines of hungry patrons in his wheelchair.

As they reached the food carts at the center of the plaza, the twins could see Solana ran two of them by herself. It definitely seemed like she could use their help.

The busy woman instructed the three to unload their fruit at one of the carts that looked like it was for making smoothies.

"My mother runs two carts every morning: a smoothie cart and a *migas* cart," Matteo explained. She will do all the preparing and cooking. Summer can take orders and payments from the customers. The three of us are going to do the serving."

"The serving?" Blaine asked.

"Yes," Matteo confirmed. "The people will order, pay, and then go find a seat at one of these tables. We will deliver their food to them."

"*Sí, hagámoslo!* Let's do it!" Summer said as she took the apron that Solana was handing to her and the twins. Blaine and Tracey had no idea their favorite local expert spoke Spanish, but they were impressed and happy that she did.

The twins nodded in understanding as the breakfast rush kicked into high gear. Summer took orders and accepted money. Solana cooked *migas* and mixed smoothies, plus pointed out where all the orders should go. Matteo, Blaine, and Tracey delivered the food and drinks. It was a little overwhelming at first, but the twins quickly got the hang of it. Matteo Perez zigged and zagged through the crowd in his electric wheelchair like he'd done it a thousand times, delivering breakfast items with a smile.

After what seemed like only minutes but was actually more than an hour, the job was finished. The breakfast crowd vacated the plaza, and the carts ceased their service. Solana Perez thanked the twins and Summer. They knew this because *gracias* was one of the few Spanish words they understood. Then, she made them eat some delicious breakfast. Afterward, she disappeared back into the alley in which she'd found them.

"Should we follow her?" Tracey asked. "Is there anything else we can do to help?"

"Thank you so much for the offer, Tracey," Matteo said gratefully. "But now, my mother will go take a rest so she can regain her energy for the next food cart rush."

The Sassafrases let out good long satisfied exhales at a job well done.

"Can I just say that your wheelchair is the coolest one I've ever seen!" Tracey said to Matteo. "I've never seen anyone move so quickly or fluidly in a chair before."

"And bro, it looks like there's even more to it!" Blaine then exclaimed.

Matteo Perez smiled and patted his wheelchair like it was his trusty steed. "Yep, I'm very thankful for my modern machine here. And for Summer and Ulysses, who helped me troubleshoot the various iterations."

"Modern machine?" The twins asked.

"That's right," Summer said. "Modern machines help humans work more efficiently and in a safer manner. They are typically built to do tasks that are too big or small or tasks that are boring, dangerous, or long. They can take slight movements or forces and

magnify them into larger movements or forces. Modern machines are usually mobile and are a combination of simple machines that are powered by gas engines, or by battery power, like Matteo's wheelchair." Summer patted the armrest of the chair.

Matteo smiled at her, adding, "Not only is my wheelchair an example of a modern machine, but so are cars, trucks, bucket dredges, airplanes, dump trucks, front-end loaders, and so many more."

"Does your wheelchair, er, modern machine, have an electrical engine?" Tracey asked.

"Well, yes," Matteo answered. "It is battery-powered, but considering batteries are essentially portable electricity storage devices that must periodically be electrically recharged, yes."

Blaine, still very excited about Matteo's wheelchair, asked, "What else can this bad boy do?"

The Mexican laughed at the boy's enthusiasm and then answered. "Well, it ha—"

A ferocious-sounding voice from across the plaza interrupted him, "Hey Matteo! You and El Fausto are going down tonight! Do you hear me, wheelchair boy? You're going down!"

The three looked over to see a big guy pointing in their direction threateningly.

"We'll see about that, Choncho," Matteo responded defiantly.

The man, who looked to be around Perez's age, scrunched up his face, changed his pointing hand into a fist and then left the plaza without saying another word.

Computer Components

After a couple long seconds of silence, the Sassafrases dared to asked, "Who was that?"

Matteo Perez laughed again. "That was Choncho Caldas."

"Oooh," Summer's eyes went wide as she leaned down and murmured, "Is he the guy?"

"Yep," Matteo replied.

"He seemed like . . . kind of . . . not nice." Tracey agreed with Summer, but more than any rude bully. She was more interested in what they'd been talking about before the interruption. She quickly cheered and turned back to Matteo and asked, "What were you saying? You know, about your wheelchair?"

"Oh yeah, I was just saying it has a computer. It's a computer that can help me do all kinds of things. Watch this!"

Matteo pressed a small button near his navigational joystick. Immediately, the left side armrest opened and out popped a flat and sleek computer monitor. The twelve-year-olds were pleasantly stunned, as was Summer, who clapped and bounced on her toes a few times.

"This touchscreen computer helps me navigate most of my wheelchair's capabilities," Matteo explained as he moved the monitor into position.

Summer shared, "Computers can be used for education, leisure, home use, and work. They are machines that store information and can perform tasks. They take in commands and spit out the designated output. Computers can be found in all kinds of everyday electronic devices."

Matteo used his fingers to work his touchscreen. Two narrow integrated treads descended from somewhere under the seat. Using the computer, he directed the tracks to plant firmly on the ground. Then, the treads lifted the whole chair up. The two bigger main wheels on each side came all the way off the ground, placing Matteo at eye level with the twins and their local expert.

"So cool!" Tracey exclaimed.

"I don't mind sitting low in my chair," the handsome Mexican said. "But sometimes it's nice to converse with my friends from a little higher vantage point."

"So cool!" Blaine repeated what his sister had just said. "It's such a cool chair, and you're such a cool guy. AND this is such great information about modern machines and computers!"

"Well, there's more where that came from," Perez smiled.

"Do you want to hear a little more about computers?" Summer asked.

"Sure!" the twelve-year-olds exclaimed.

"Wow, the two of you might actually like science as much as I do," Matteo chuckled.

"Possibly," the twins said.

"Most computers are made up of the following parts," the local expert started again. "The screen, the keyboard, and the CPU. Screens, also known as monitors, display words and images. Keyboards type in words or input instructions. This can be done on a traditional keyboard, you know, the one with the letter, number, and symbol-covered buttons. A computer mouse is another way to input instructions. And then you have a touchscreen like I have here, which combines the components of a screen and a keyboard."

The siblings nodded in understanding. Then Tracey asked, "And what about the CPU? What exactly is that?"

"The CPU acts like the brains of the computer," Summer

explained. "The CPU contains the ports, such as USB ports, which can transfer information. It also contains the battery, to store and provide the electricity the computer needs to work. It contains the processor to perform the calculations that make the computer operate. It has the motherboard to connect the different parts of the computer to let them communicate. There's the RAM to store information when the computer is on, and there's a hard drive to store information even when the computer is off."

"Whoa, that's a lot," Blaine exhaled, and a quizzical look settled on his face. "Wait, wait, what's the difference between a hard drive and hardware?"

Matteo clarified, "A hard drive is a component of the computer while the hardware is the computer."

Summer could see on Blaine's face that he didn't fully get it, so she expanded on Matteo's explanation, "Hardware is a physical device, like a computer or a smartphone. Then software is the set of instructions in bits and bytes that tell the hardware what to do."

"Bits and bytes?" This time it was Tracey asking the question.

Summer nodded. "A bit is one number in a computer code, while a byte is an 8-bit group that carries information. A kilobyte is 8,000 bytes. A megabyte is 8,000,000 bytes. A gigabyte is 8,000,000,000 bytes, and a terabyte is 8,000,000,000,000 bytes!"

"Eight trillion?" Tracey asked with wide eyes.

The local expert nodded.

"That's a lot of bytes," Blaine affirmed.

Almost time. He'd had the extra-large Forget-O-Nator set up as a classroom for a couple of days now. He'd sent out the junior high class reunion invitations about a week ago. After all his attempts at sabotaging the twins' science-learning had failed, this was his last resort for exacting revenge on Cecil Sassafras.

He'd been hurting and grieving the loss of what Cecil had taken from him for decades! The Bunsen burner exploding in his face. Losing his beloved eyebrows all those years ago. Now finally, after all this eybrowless time, vengeance would be served. He would get everybody in the Forget-O-Nator and wipe it all away.

Everything.

"It can even go up the stairs!" Tracey exclaimed, almost not believing what she was seeing.

Matteo Perez chuckled and nodded proudly. He'd shown Blaine and Tracey a lot of what his wheelchair could do, but he'd saved the best for last. The integrated treads, which could extend out to about the length of a pair of skis, indeed could take his chair both up and down flights of stairs. The Sassafrases were truly awed by this modern machine and wonder of physics.

The stairs Matteo and his chair currently climbed were an offshoot of the plaza the three had served breakfast in. Once they reached the top, he transformed his wheelchair back to the mode it had been in when the twins had first met him. Summer clapped her hands and did a mini-dance at the top of the Saltillo-tiled stairs.

"Hey, so what was the deal with that Choncy guy?" Blaine asked, shifting gears a bit as the group walked on.

"You mean Choncho?" Matteo asked.

"Yeah," Blaine confirmed. "Him. Why was he so mean to you? What did he mean when he said you were going down tonight?"

"He's talking about *Lucha Libre.*"

"*Lucha Libre?* What's that?" Tracey asked.

"Wrestling," Perez responded. "I'm going to be wrestling against him tonight."

This was not what the twins thought Matteo was going to say.

"That's where we're going right now," the optimistic young man said. "To get everything ready."

"You're going to wrestle against that huge guy?" Tracey asked.

Perez nodded. The twins looked at Summer, whose face displayed a wide grin with no trace of concern. Their heads swiveled back to Matteo.

"Matteo, you're in a wheelchair," Blaine said. "How are you going to wrestle, Cho-Cho?"

"Choncho?"

"Yes, him. I mean, your wheelchair is one of the coolest things I've ever seen, but how will you wrestle with that guy?"

Matteo pushed his joystick forward and started moving again. The twins walking on one side, Summer on the other.

"I've wanted to be a *luchador* for as long as I can remember," Perez said, that smile of his still present on his face.

"Doesn't *luchador* mean wrestler?" Tracey asked.

The young man nodded. "I would wake up early on Saturday mornings to watch *Lucha Libre* on TV. Oh man, those guys were great! Jumping and flying all over the ring. I watched all the guys battle and became a huge fan! I might've even been the biggest fan of *Lucha Libre* in all of Mexico! I spent hours studying the moves and then came up with a whole series of my own moves. I even made my own luchador masks with materials I found around the house. I just dreamed of the day that I could compete in the ring myself!" Matteo smiled his biggest smile yet.

"My mother didn't like *Lucha Libre*, not one bit," the young man laughed. "She thought all that wrestling stuff was dangerous. She thought I could get hurt if I joined in. She did not want me to become a wrestler. She made that very clear."

"Is that what happened?" Tracey asked compassionately. "Is that why you're in a wheelchair?"

Perez brought his wheelchair to a stop. He looked at Tracey and shook his head. "Nope. It was a car accident. Spinal cord injury to my thoracic vertebrae. I lost the use of my legs."

Summer rested a hand on Matteo's shoulders, her normal, bubbly manner somewhat subdued as they walked.

"I'm so sorry, Matteo," Tracey said, tears springing to her eyes as she listened.

"I was pretty down about it for a while," the young man admitted. "Thought maybe my life was over. But it wasn't. Summer and Ulysses taught me it was just going to look a little different."

Summer squeezed his shoulder and let go.

"I also have my wonderful mother to thank for not letting me wallow in pity," Perez continued. "She got me my first wheelchair and made me come to work with her at the food carts. And I'll be honest, it was a little rough at first. I had a real hard time trying to serve food while navigating my wheelchair by arm strength."

"You were serving food and pushing your wheelchair by hand?" Blaine asked, amazed.

"Well, I was at least trying," Matteo laughed. "But it was very frustrating. I would often drop orders and bump into people. The customers tried not to get frustrated or stare, but many of them couldn't help it. I knew if I was going to keep helping my mom, something needed to change."

"So what'd you do?" Tracey asked.

"Well, as I said before, like the two of you, I had always loved science and specifically, physics."

The Sassafrases nodded. They did indeed love science.

"So I put that science to use as I set out to design a better wheelchair for myself."

The twins nodded again.

"I needed an electric chair that could navigate the bustling crowds at the plaza better. One with a tight turning radius. One with some kind of surface for carrying food and drinks. There were already some good cutting-edge wheelchair designs out there, but they were all very expensive. Certainly more than my mother and I could afford. So we couldn't purchase anything like that. But I really wanted to design the chair myself, anyway. You know, make it custom."

"So, is this the chair you came up with?" Blaine asked.

Matteo chuckled kindly again. "This is . . . I think . . . the twenty-first iteration. In the process, I ran across Verano Playa and El Asistente in a robotics forum. Summer and Ulysses helped me tweak and adjust my prototype and we became friends along the way."

Summer clapped her hands rapid-fire style as she beamed at Matteo.

The twins now understood a little more about his story, his mom, his accident, and his connection to Summer. Plus, his love of physics and robotics. But what about this whole *luchador* thing? Was he really going to wrestle that big guy who threatened him? Even the possibility of this worried the Sassafrases.

"So? What do you say?" Matteo beckoned his friends as he surged his wheelchair forward once more. "Is everyone ready to go to the wrestling ring with me?"

Chapter 17: El Lucha Libre

Coding Cyborgs

"It's a GPS," Matteo Perez said with a smile as he faced the sleek touchscreen computer. "Which stands for Global Positioning Satellites. It's a network of thirty-one satellites that can determine your position anywhere in the world." Matteo punched in the address for a place called *Plaza del Partido* and the chair moved on its own.

"Your wheelchair has a GPS! That's so awesome!" Blaine exclaimed as he looked at the map on the computer screen, pinpointing where they currently were and where they wanted to go, drawing a route between the two points.

"Ulysses provided Matteo with the program to access the Global Positioning Satellites," Summer said. "But it was Matteo who figured out how to link the information into his mobility system. It's all so exciting!"

The group of four meandered through the colorful streets of Guadalajara, taking the rights, lefts, and straightaways as the map directed. Matteo mentioned earlier about people staring at him, but out here on this sunshiny route today, there didn't seem to be any stares. There were only smiles and waves from the people they passed. Maybe that was because Matteo was being so friendly as he went. The handsome young man was smiling and waving at virtually everyone.

As they continued to walk the picturesque streets. Blaine gave his new friend another compliment, but also asked the worrisome question rolling around in his mind. "Matteo, you're awesome. Your wheelchair is awesome. You've taught us a lot about physics, but . . . I mean . . . are you really going to wrestle Chester?"

"Choncho?"

"Yes. Him. Are you really going to wrestle that guy tonight? I mean, couldn't he really hurt you?'

"Oh, I'm not just going to wrestle him," Perez said, ever smiling. "It's *Lucha Libre*, which means freestyle wrestling. There will be lots of wrestlers in the ring all at the same time!"

The twins gulped. Now they were even more worried.

A few more hours. That's how much longer he would have to live with these memories.

He rubbed his hairless brow, remembering buying the house two doors down from Cecil so he could monitor all that his red-headed archenemy did. He'd watched as Cecil excelled in virtually everything, becoming a pseudo-famous research scientist. The memory of Cecil and his prairie dog lab assistant coming up with the apex invention of the invisible zip lines now danced in his brain. The image of him stealing and copying the concept so he could travel on invisible zip lines too flitted across his mind.

He had planned to sabotage those twins' science learning in order to crush Cecil's dreams. He'd stalked those twins and marooned them in the wilds of Africa, trapped them in a tomb in Egypt, and spied on them with a robot hummingbird. He had scrambled their SCIDAT data, sending them to two locations and trapped them in virtually indestructible boxes. He'd pretended to be a male nurse and attacked them with robotic squirrels. Yet, those twins wouldn't stop zipping around, learning science. His desired revenge remained elusive, and his eyebrows still hadn't grown back.

So he stole the Dark Cape suit and the plans for the original model of the Forget-O-Nator. He wiped Summer Beach's memory, but Cecil restored it. He kidnapped villains and wiped their memories, intending to use them against the Sassafrases.

Instead, he had fallen in love with a former Swiss Secret Service agent, Adrianna Archer. Together, they'd commanded an immense army of servile scientists from an immaculate lab in Siberia. He'd created an army of Dark Capes and an extra-large Forget-O-Nator. Then he'd lost Adrianna, his Siberian lab, and his army of Dark Capes. But none of it worked.

All he had now was his Dark Cape suit, his extra-large Forget-O-Nator, and one last remaining plan for revenge against Cecil Sassafras and all that he loved.

He had returned to his home in the old neighborhood, hoping it was a safe place to hide, despite the fact that Cecil and Summer had pinpointed this spot as his base of operations. He'd sent out invitations to a fake junior high class reunion and renovated the large Forget-O-Nator to look like part of the junior high building. He'd waited and waited and waited . . .

Enough!

Thaddeus Wazeezy stood up from his seat, reached his fists high in the air and shouted in rage. "Enough! Enough waiting! Enough having a hairless brow!"

It was time for his revenge to come to pass. He wasn't just going to stop Cecil, he was going to stop those twins. And he wasn't just going to stop those twins from learning science. He was going to wipe away everything they'd learned: forever.

Yes, his own memories would disappear along with Cecil's and that niece and nephew of his. That was just fine by him. Everything was about to be destroyed, deleted, and wiped away.

His revenge would finally be complete. It was almost time.

"Wait . . . wait . . . what?"

Part of learning and adventuring was being amazed. The twins had found this to be true over the course of the summer. Again, they found themselves shocked to see something they had not at all expected.

The GPS had successfully led them to the *Plaza del Partido*. They had found a wrestling ring with a raised blue platform and red, white, and green horizontal ropes. They had also found the *luchadores* who would wrestle in the ring later this evening. The twins' mouths dropped in shock. The *luchadores* were not . . . human—they were . . . robots!

"Tada!" Summer held out her arms wide in introduction.

"Wait . . . so . . . huh?" Blaine bumbled.

"So this means . . . you're not . . .. umm," Tracey fumbled.

Matteo Perez looked at the twins in realization. "Oh, now I see," he laughed. "When I said I was going to wrestle tonight, you thought I meant I myself was going to physically get in the ring and battle."

The Sassafrases nodded.

"Well, when I was younger, that's definitely what I wanted to do," Perez said. "As I designed my wheelchair, I fell in love with computers, coding, and robotics. My dream to become a *luchador* remained, but it changed from becoming a traditional *luchador* to becoming a designer, programmer, and coach for a robotic *luchador*."

The twins were more than a little impressed as they took another wide look around *Plaza del Partido*. Overhead, the plaza was topped by several strings of brightly colored flags flapping in the breeze. The ring in the middle of the plaza was surrounded by rows of wooden bleachers. The edges of the plaza were lined with large booths that looked like lockers or storage units. Most of them were open, revealing large robots clearly designed to look like Mexican *luchadores*. There was a lot of variety, but all resembled humans wearing their own signature *luchador* masks. The average height had to be seven feet tall!

"Ulysses!" Summer cried out. The twins spotted the Arctic Ground Squirrel, standing there next to a muscled metal robot with a mask in the same colors as the Mexican flag.

Blaine was about to ask if the robot next to Ulysses was Matteo's, but before the boy got his question out, another cheerful voice burst onto the scene.

"Matteo! What's up, *mi amigo*? Are you ready to win that cash prize?"

"Hey, what's up, Gordo," Perez responded to the approaching jovial, heavyset man with a smile and a fist bump.

"Nah, *amigo*. You know I do this for the love of the *Lucha Libre*, not the *pesos*."

The big guy nodded like this was actually true for him as

well.

"Blaine and Tracey, this is Gordo Gomez. He's the designer, programmer, and coach of a robot *luchador* by the name of *El Burrito*! Gordo, this is Blaine and Tracey Sassafras. They are hanging out with Summer and I today talking science."

"Super cool!" Gordo responded enthusiastically, then said, "No one else is here yet except for Gonzalo, and I already have El Burrito programmed and coded like I want him, so I'm going to grab some lunch. Can I get you guys anything?"

"Lunch?" Matteo laughed. "We just had breakfast not too long ago."

"Oh, c'mon, amigos. I'm always hangry."

Both Mexicans laughed.

"Alright then. I'll see you guys later," Gomez said and then quickly disappeared from the square.

Matteo waved to the only other person in the square. The older man sat on a stool at the other side of the plaza. He didn't wave back, but he did give a wry smile.

"That's Gonzalo Viejo," Perez told the twins. "He was a *luchador* himself back in his younger days. He went by the name *El Grunidor*, which means the Growler. Tonight, he will serve as the referee for our *Lucha Libre* match. Right now, he is guarding all the robots."

To the twins, Gonzo the Growler looked like someone not to be messed with. Matteo guided his wheelchair around the ring toward Ulysses and the robot he was perched next to. The sleek robot was probably the most humanlike in the bunch. It had contoured human muscles made of some kind of metal. They didn't see any visible gadgets or accouterments like some of the other robots. The machine wore a mask with the colors of the Mexican flag: red, white, and green. It stood still and silent in front of its locker.

"Blaine and Tracey, meet El Fausto!" Matteo said with outstretched arms.

"You made this?" Blaine exclaimed in flabbergasted amazement.

"Isn't it amazing?" Summer clapped.

"It is!" Blaine exclaimed.

"Thanks *amigos,*" Matteo responded. "I took all that I learned from designing wheelchairs and put that knowledge into this guy. Most of the other robots have some kind of signature move or intimidating weaponry, but *El Fausto* looks more like a traditional *luchador* and competes like a mixed martial artist." Matteo grinned.

"Wow! That sounds even better than REESE!" Tracey exclaimed.

Forgetting for a second the designer of REESE was with them. Even so, Ulysses S. Grant didn't look offended at all. Yes, he had created the robot to help the twelve-year-olds with their summer learning, but right now the Arctic ground squirrel just seemed excited for Matteo.

"What do you mean by saying this robot competes like a mixed martial artist?" Blaine asked.

"El Fausto can be programmed to operate in the ring using moves from the leading martial art disciplines, whether that be wrestling, Karate, Judo, Ju-Jitsu, boxing, or some other martial art. I usually program him based on who he will face in the ring."

Blaine sort of understood this explanation. "So is *El Fausto* all ready to go for *Lucha Libre* tonight?"

"Actually no," Perez said. "I wanted to come here early today to make a couple of changes and to rewrite some of the code with the help of Summer and Ulysses."

"The code?" The twins asked.

"Yep," the young man answered and then pointed at his

luchador robot. "*El Fausto* is basically just a big computer."

Summer wiggled her fingers and said in a giddy science-loving voice, "The instructions that tell a computer what to do are known as code and writing them is called coding. These codes are a combination of symbols and words."

"Coding has both input and output. Input is a set of instructions that, when received, tell the computer to specifically show or do something. What the computer then shows or does is the output." The Sassafras twins recognized Summer was making sure they got all the SCIDAT information they needed.

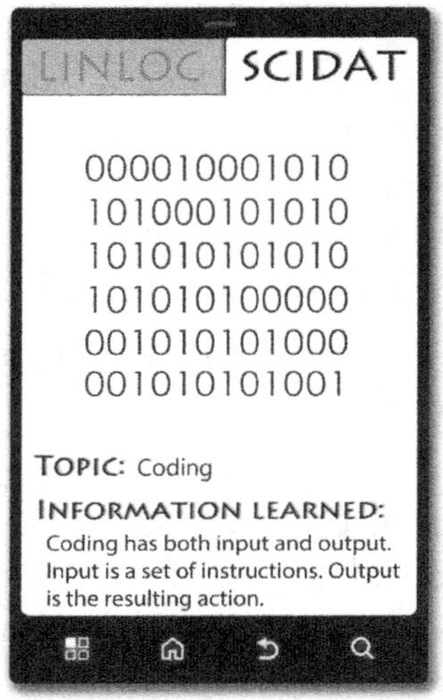

"Coding is done by people known as programmers," she continued. "And there are many different coding languages. Some examples are Python, Scratch, Java, C++, and more."

"Those are languages?" the learners asked.

"I know. Crazy, right?" Matteo answered. "We use C++ and Python with *El Fausto* because those are considered to be the best languages for robotics. Some coding languages let you drag-and-drop blocks of code, while others require the programmer to type out the lines of code. Using both allows us to have speed and efficiency in managing memory, plus readability of extensive libraries. It does really well in its interactions with robot sensors and actuators, making it ideal for precise control of the robot and for high performance."

"What does the robot's name mean?" Blaine cut in with a question.

Their local expert smiled. "*El Fausto* means the Stupendous!" Summer answered excitedly for her Mexican friend.

Matteo nodded to confirm the female scientist's translation. "Super cool."

Blaine responded, and Tracey agreed.

"Would you like me to bring him to life?" Perez now asked with a sparkle in his eyes.

"Sure!" the brother and sister exclaimed.

"I've switched off the GPS here on my computer, and I'm going to switch to a program that allows me to wirelessly connect to my robot using Bluetooth. This will also give us the access we need to do some coding."

As the computer successfully connected to the robot, *El Fausto* suddenly became illuminated with fine lines of green, red, and white light, organized in accentuating designs running from neck to feet. His body stayed still, but his head moved slightly and his eyes lit up as bright orbs.

"*¿Qué pasa mi entrenador?*" *El Fausto* said, clear and bold.

The Sassafrases flinched, but it was surprise and amazement that made them do so, not fear. "What did he say?" Tracey asked.

"He said, 'What's up, my coach?'" Matteo smiled. "That means we are successfully connected."

Ulysses hopped up on the arm of the wheelchair and Summer leaned in. All kinds of data could now be seen on the wheelchair's computer screen.

"Looks like he's fully charged and functional. All systems are active and normal, and we are now free to begin coding."

"Which martial art are you going to program him with now?" Blaine asked

"How about Krav Maga for tonight?"

"Krav Maga? What's that?" Blaine asked.

"It's a martial art that was developed in Israel and is widely considered the best martial art for defense. I think there are going to be at least eight robot *luchadores* in the ring tonight, some who are very aggressive. So I think *El Fausto* should play it quick and defensively."

"There are going to be eight robots in there at the same time?" Blaine asked. "So how do you win? I mean, if there are eight robots all fighting at the same time, won't it be hard to tell who the champion is?"

"It's the last one standing," Perez responded.

"The last one standing?" Tracey repeated.

"Yep, the last robot in the ring will be the winner of the match," Matteo shrugged. "What can I say? It's *Lucha Libre*—freestyle fighting."

"And that last one standing is going to be *El Fausto*!" Summer exclaimed with undaunted belief in her friend and his robot.

Robotic Riot

Blaine asked another question, "Will there be a crowd here tonight?"

Matteo shrugged. "Maybe. *Luchadores* have been popular in Mexico forever, but robotic *luchadores* are a new thing. So not many people know what we are doing, but those of us who are in it love it. All that said, we do have some very dedicated fans. We are a small but pretty tight-knit community. And as you can see, people travel from all over the world to take part."

Summer and Ulysses nodded.

Blaine nodded, still wondering about the little run-in they'd had this morning. "What about Churro?"

"Choncho?"

"Yes. Him. What was his deal? Why was he so rude to you earlier?"

"Oh, I don't know," Matteo said. "Choncho has always been a little bit like that. Really, he's not the threat. I've beaten him before. The real threat is a *gringo* by the name of Kevin Maye."

"A *gringo*?" Blaine asked.

"Kevin Maye?" Tracey added.

Matteo nodded. "He's a wealthy guy from Texas. I think his mother was Mexican and his dad was American, but he has blonde hair, bright white teeth, and a cocky attitude. His robot is called *El Texican*. He doesn't store it here. He'll bring it in on a big fancy trailer."

Ulysses chattered loudly, and Summer agreed.

"You're right! He's probably spent ten times the amount of money on that thing than any of the rest of us did on our robots."

"Like I said, this whole robot *luchador* thing is kind of new. Most of us don't have a lot of money. We mainly do it for the fun of it, even though there is usually a small cash prize for winning, which is obviously very helpful. But Kevin came in here and upped the ante. *El Texican* has spurs that can stab through most metals and a belt buckle that can shoot out like a bullet and return to its place. Plus, a cowboy hat that spins like a diamond blade saw. Cuts like one too. I'm kind of afraid he's going to destroy all our bots tonight."

"Well, aren't there any rules?" Tracey asked. "Can he really just go around and destroy the other robots? What about Gonzo the Growler? Can't he stop Kevin and *El Texican* from doing that?"

"Like I said," Matteo shrugged. "The whole robot *luchador* thing is kind of new, and there really aren't a lot of rules yet."

Blaine looked around at all the other robots in the nearby lockers. Then he looked back at *El Fausto*. "I think your robot is going to win, Matteo. I like *El Fausto* the best. *El Texican* sounds intimidating, but I bet you can beat him. Yep, *El Fausto* is the best."

Tracey agreed. "You guys did such a good job in designing, building, and coding *El Fausto*. He looks the most like a real human *luchador*. I like that."

"Those are really kind words. Thanks, Sassafrases," Matteo Perez accepted the compliment graciously.

Summer piped up. "*El Fausto* is an android. Really, all the *luchadores* here are androids."

"Androids?" Blaine asked.

"Yep. Androids are robots that have designs based on the human body. Matteo designed *El Fausto* like he did because of his love for wrestling. The more humanlike the robot is, the more capable it would be to perform true wrestling moves and other martial arts. Robots should perform the tasks they are designed and programmed to do."

"The first robots were built in the 1980s. But all are machines built to do work on their own. Each has its own unique look based on the task it's designed to do. Robots in factories are typically strong and able to do consistent, repetitive tasks. Robots in the medical field can perform very delicate tasks. Robots in space can do tasks that are too dangerous for humans to do. *El Fausto* is designed to be a *luchador*. See what I mean?"

The twins saw what their favorite local expert meant. They again thought about REESE, their robot friend that Ulysses S. Grant had designed to help them with exploration, entertainment, and scientific enhancement. They wondered how he would do here in Mexico in a robot wrestling match. Probably pretty well, they thought.

"Robots are usually powered by electricity," Summer continued. "Many have hydraulics to move their mechanical limbs. Basically, all have computers for brains. Robots can see and feel, but of course not in the same way humans do. They have sensors which are coded to interpret the information received."

"Does *El Fausto* have hydraulics? Or any of those cool stair-climbing treads like you have in your wheelchair?" Blaine asked. "He's so sleek it's hard to tell all the things he might be capable of."

Perez smiled. "Let's just say *El Fausto* may have some surprises up his sleeve."

"We want *El Fausto* to beat Kevin Maye and *El Texican*. And we also want you to beat that rude Chacha guy," Blaine said, determined.

"Choncho?"

"Yes. Him. We really want you to beat him!"

"Yes, we really want you to win tonight," Tracey agreed.

"Cool, we've seen Ulysses' robot surprises before, so we can't wait to see what you all came up with," Blaine said.

They didn't want to hurt him. They just wanted to wake him up. Both literally and figuratively.

Right now, he was asleep in the middle of the day up in his room. But even when he was awake, he wasn't himself. He was just moping around with no energy, excitement, or eccentricity. It wasn't right, and President Lincoln knew it. He knew he needed to do something about it, so the prairie dog had come up with a plan.

Cecil thought he'd missed his chance with Summer. He thought Summer was in love with Paul Sims. The Prez didn't know how Cecil could be so blind to the truth. Summer didn't love Paul! She loved Cecil! And she always had. Just like he'd always loved her.

He needed to wake up, realize this was true, and go fight for her! Right now, President Lincoln was going up the cluttered stairs with REESE. The robot used his strong mechanical arms to ascend, following closely behind him.

They made it to the upstairs hallway and then to Cecil's bedroom door.

They pushed through the door. There was Cecil, asleep in his bed, face down, arms by his side, rump in the air, and messy lab coat and bunny house slippers still on. There was some slobber, snoring, and, of course, some messy red hair.

Without pause, the two took their places: Lincoln on the bed next to Cecil, and REESE by the door. The two looked at each other and then President Lincoln began a silent countdown. At zero, the kind-hearted mayhem began. A flurry of fury slaps rained down on Cecil's face. The room suddenly filled with light from the robot's monitor accompanied by loud music—music that could be described as disco-mantic—a blend of disco and romantic music.

The Sassafras scientist awoke with a start and a look of confusion. "Welly-wee-willikers! Sassy Molassey! What in the wide world?"

President Lincoln stilled, hoping the plan would continue to play out. And it did. Cecil's attention was grabbed by the music REESE played and the images flashing on the robot's torso screen. That was good. The pictures were of Cecil Sassafras and Summer Beach together over the years, from junior high through the last time they'd bowled together at the Ambidextrous Octopus.

President Lincoln had worked hard to compile all these

images, and it was working. Linc-Dawg could tell by the look on his favorite human's face, this was hitting deep. Realization resonated in the man's heart. Determination stirred in his soul. This was going to do the trick. President Lincoln just knew it.

Cecil's eyes watered as he watched the montage. When the video finished and the music stopped, he paused for a quarter of a second. Then Cecil bolted to his feet right there in his bed.

"I love you, Summer Thyme Beach! And I'm going to fight for your heart!"

The *Plaza del Partido* was full of people. The bleachers around the ring were full, and people stood behind the bleachers waiting for the festivities to begin. Evidently, robotic wrestling was a little more popular than Matteo had made it out to be. Gonzalo, aka Gonzo the Growler, was dressed in his black and white stripes and stationed ringside, ready to officiate. All the robots were alive and activated! The designers, programmers, and coaches of the robotic *luchadores* were ready. Except one. Kevin Maye and *El Texican* had not arrived yet. Summer, Ulysses, Blaine, and Tracey stood in the back between Matteo Perez and *El Fausto* and *El Burrito's* team with Gordo and his sister.

El Burrito had the form of a human and a burrito mixed together. His dominant power, according to Gordo, was the ability to wrap other robots up and squeeze them until their programmers tapped out. Even now, as they waited, Gordo gave the Sassafrases a rundown of all the other wrestling robots.

"Okay, that one right there with all the spikes poking out everywhere. That's *El Hombre del Cactus*, which means The Cactus Man. Its dominant power is puncturing other robots with all those sharp spikes."

It wasn't hard for the twins to look at *El Hombre del Cactus* and think of their own run in with the infamous Cactus Head Monster earlier in the summer. They sure hoped *El Fausto* could avoid all the cactus robot's protrusions.

"And that one over there that has arms that are way bigger than its body," Gordo said, pointing. "That's *El Largo*. His dominant power is grabbing other robots, picking them up, and then throwing them clear out of the ring. Oh! The pink robot with all those long, rope-like things attached to her head. That's *Las Trenzas Rosas*. Her dominant power is entangling other robots up in all those super strong cords and therefore rendering them ineffective."

"Okay, way over there on the other side, the robot that looks kind of like a wrestler and a rocket mixed together, that is *El Cohete Loco*, aka The Crazy Rocket. His dominant power is blasts of ramming and pushing power. His programmer happens to be the nephew of an actual Mexican *Luchador.*"

The twins looked at each other with wide eyes. They knew the *luchador* Gordo was talking about. They'd met him in New Zealand while studying anatomy. He had been a participant in the PGC, the Professional Gamer Championship.

"What about Cosco?" Blaine asked.

"Choncho?"

"Yes. Him. What's the name of his robot?"

"*El Caliente Picante,*" Gomez answered. "Which is basically like saying Hot and Spicy. It's the robot that's completely red. Its dominant power is shooting out sparks and fire, which a lot of us think should not be allowed. But hey, this robotic *Lucha Libre* is still kind of like the wild west, right?" Gordo Gomez chuckled and shrugged his shoulders.

The Sassafrases, however, couldn't find it in themselves to laugh. They were getting too nervous about this whole *Lucha Libre* thing. Who was going to be the last robot standing? They really wanted *El Fausto* to win, but all these other robots he was going to go against sounded scary.

Summer seemed nervous too, which wasn't like her. Even Ulysses was taking anxious breaths. The two truly were invested in this robotic *luchador* thing with their friend Matteo Perez.

He was so sure of himself. He was confident his plan had worked. He was sure of it. Summer Beach was falling in love with him.

Their date had gone splendidly. She loved the time they'd spent together. She hung on every word he said. Sure, he didn't love her. Not at all. But that didn't matter. What mattered is that Triple S thought she loved him.

Whenever they showed up, and he had a feeling that would happen soon, Summer was going to be so in love with him that there would be no way these Swiss Secret Service agents would arrest him or charge him with the crimes he'd committed. He was sure of it. But where had Summer run off to?

Just like Matteo had said he would, Kevin Maye had arrived at the *Plaza del Partido* with a bunch of hoopla. The *gringo* and his robot *luchador* had pulled up on an overly-decorated trailer pulled by an expensive-looking truck with loud music blaring. Kevin and *El Texican* alighted from the trailer and entered the plaza like they had already won the wrestling match.

Blaine and Tracey shuddered as all the robots stood in the

ring, ready for *Lucha Libre* to start. The twelve-year-olds were scared for *El Fausto*, and they were scared for Matteo.

Gonzo the Growler climbed into the ring. All the programmers had their computers, tablets, or phones ready.

Matteo Perez sat in his wheelchair with his touchscreen computer ready. He looked as relaxed as he'd looked all day. "How can our friend be so calm?" The Sassafrases wondered nervously.

In the ring, *El Texican* was bigger and taller than the rest, his spurs, belt buckle, and cowboy hat shining and ready. *El Caliente Picante* looked red and angry and ready to set something on fire. *El Hombre del Cactus* looked ready to poke, prod, and puncture. *El Largo's* huge robotic arms looked more than ready to grab, lift, and toss. *Las Trenzas Rosas'* strands and cords of robotic "hair" looked ready to entangle and incapacitate. *El Cohote Loco* looked ready to shove. *El Burrito* looked ready to squeeze. Then there was *El Fausto*, with his robotic muscles and cool lines of light shining, looking as calm as his creator.

The gathered crowd quieted down, as if anticipating the start of the match.

The Sassafrases gulped.

Gonzo lifted his hand in the air.

In a flash, the Growler's hand came down.

"Fight!" He shouted.

Mayhem. Pure mayhem filled the wrestling ring in *Plaza del Partido* after the drop of Gonzo the Growler's hand. The programmers tapped furiously at their screens. The eight robot *luchadores* went ballistic.

El Fausto immediately employed the best of Krav Maga defensive method as he fielded attacks from several of the other robots. Meanwhile, *Las Trenzas Rosas*, the pink robot with electric dreadlocks, surprisingly charged directly for *El Texican*. Before the big expensive robot could kick up his spurs or shoot his belt buckle or get his hat to spinning, he was wrapped up by strong mechanical chords. *Las Trenzas Rosas'* entangling strands had rendered *El Texican* obsolete.

However, before any celebration could happen, *El Cohete Loco* came rocketing across the ring directly toward the two and hit both robots with such force that they flew through the ropes and out of the ring.

At the same time this was happening, the other four robots did their best to avoid the advances of *El Hombre del Cactus.* The spiky robot was slow, but relentless. *El Burrito* moved to its left. *El Largo* moved to its right. *El Caliente Picante* shot a couple of sparks of fire at the attacking cactus but missed.

To the crowd's surprise, *El Fausto* showcased an unexpected skill as he sprung up and jumped over the top of the barbed hombre. The pursuing cactus then made a surprisingly quick jaunt to its left, finding *El Burrito. El Hombre del Cactus* directed all its sinister spikes toward Gordo's robot, but it was immediately wrapped up by the large mechanical flaps of the one it had attacked. *El Burrito* then gave a mighty squeeze. *El Hombre del Cactus* was crushed.

Gordo wanted to celebrate, but the big Mexican quickly realized that while he had taken out *El Hombre del Cactus* with his squeeze, the other robot had simultaneously taken him out with his robot spikes. Both robots were down for the count and were quickly picked up and thrown out of the ring by the powerful arms of *El Largo.*

El Caliente Picante made a quick move toward *El Cohete Loco* and took a shot toward the rocket robot. The burst of flame somehow found its way through the small cracks of the robot's exterior to land inside the machine. There was only a brief pause and then, to the shock of all, *El Cohete Loco* exploded. Some pieces of the robot remained in the ring, but most of the pieces flew out and landed dangerously close to the crowd.

Meanwhile, *El Largo* and *El Fausto* were engaged in hand-to-hand combat that looked more like actual wrestling than anything that had been displayed in the ring so far. *El Largo* was using more of a wrestling approach, while *El Fausto* continued to operate according to his Krav Maga coding. Matteo's robot proved to be more fluid, flexible, and humanlike than *El Largo*. He not only successfully eluded *El Largo's* attacks but also used gravity, torque, and momentum to toss the robot out of the ring.

Now there were only two robots left: *El Fausto* and *El Caliente Picante*. The Sassafrases looked over toward Choncho Caldas. The big guy was staring daggers at Matteo Perez. Choncho had called Matteo out this morning near the food carts. He had shouted with malicious certainty that Matteo was going down. Now it was time to prove it.

The remaining programmers brought their attention to the computers in front of them and tapped in the next directives.

El Caliente Picante did something he had not done yet—he glowed red, redder than he had been up to that point. The fire inside became visible. He was going to blast *El Fausto* with a ball of fire! Choncho must have been saving this just for Matteo's robot. Blaine, Tracey, Summer, and Ulysses cringed, unsure the machine could defend against a flame.

As *El Caliente Picante's* flame continued to build, *El Fausto* also did something he had not yet done. He sat down. "Oh no!" The Sassafras' minds screamed. "Is Matteo giving up?"

But then they saw it—out of *El Fausto's* hips came wheels, one from each leg. They were big wheels, like those of a wheelchair.

Then, before any flame could be flung, *El Fausto* wheeled toward the other remaining robot and made contact, knocking the red robot up into the air. *El Caliente Picante* then came back down, landing on *El Fausto* almost like a person landing in a wheelchair. Without pause, *El Fausto* wheeled *El Caliente Picante* to the edge of the ring and dumped him off the platform. *El Caliente Picante* landed on the ground with a metallic thud, fireball unspent.

Only one robot *luchador* now stood in the ring. Matteo Perez and his versatile robot had won *Lucha Libre*. Gonzo the Growler came over and lifted *El Fausto's* arm to seal the victory. The crowd erupted in applause.

CHAPTER 18: THE FINAL ZIP

Bonus Data

It was probably the largest happy jumping dance hug there had ever been. Upon the victory of Matteo Perez and his robot *El Fausto*, Summer Beach had pulled virtually the entire crowd into a celebratory mobile hug. Even Choncho Caldas, who in defeat added a sincere apology for how he'd treated Matteo, joined in.

Eventually the celebration had died down. The crowd, the programmers, and even Gonzo the Growler had left the plaza. Now only four remained. They'd said goodbye to the Perez's and now Blaine, Tracey, Summer, and Ulysses sat facing each other at the center of the *luchador* ring. All the SCIDAT was entered. The pictures were sent in.

Currently, the Sassafras's were opening up their LINLOC

apps to view where they would be zipping to next. Summer Beach seemed to already know about the next destination, and it was all she could do to contain herself.

"Uncle Cecil's basement!" The twins exclaimed as they saw it at the same time.

"Yes!" Summer shouted as she exploded up to her feet. "Yippity skippity yip yip yooray! You two wonderful little cutie-frasses have done it! Not only have you successfully completed the study of physics, you've successfully completed your entire summer of learning!"

"We've done it?" Tracey asked shocked.

"Oh my cow! We've done it!" Blaine blurted, also in a slight state of shock.

"You sure have, you two amazing smartie-frasses!" Summer beamed. "Now get out those harnesses, helmets, and carabiners of yours! Ulysses and I will do the same. Then, let's zip!"

The twins exuberantly obliged, and in a matter of seconds, found themselves hanging in the air alongside their favorite local expert and their favorite arctic ground squirrel, waiting to shoot off onto the invisible ziplines.

Maybe for the last time.

They were thrilled about their monumental accomplishment, but there were tiny tinges of sadness. Because if this was indeed their last time to soar across the invisible ziplines their uncle had invented, they sure were going to miss it.

"Here we go!" Summer shouted just as the four took off into fantastical swirls of light.

This time, the zip lining seemed to happen in slow motion instead of at light speed. It's as if time wanted to slow down enough for Blaine and Tracey to enjoy this last ride. They felt the gushing of wind in their hair. They reached out and touched the moving light with their fingers. They felt the embrace of the opportunity they'd been given in their hearts.

Way too soon the zip lining ended and Blaine, Tracey, Summer, and Ulysses landed with a jerk, their carabiners automatically unclipping from the invisible lines. Even before their sight and strength returned, the twins could hear Summer Beach squealing in delight over what they'd accomplished.

When sight and strength did return, Summer was still squealing, and the twins saw that their uncle's basement was back to looking like it had been when they had first seen it. Which was to say: cluttered, jumbled, and awesome.

All at once, the brother and sister felt their phones vibrate inside their backpacks. They both immediately recognized this as a sign that Bonus Data had arrived on their devices. Bonus data was a little gift of scientific information they received for successfully completing a scientific subject.

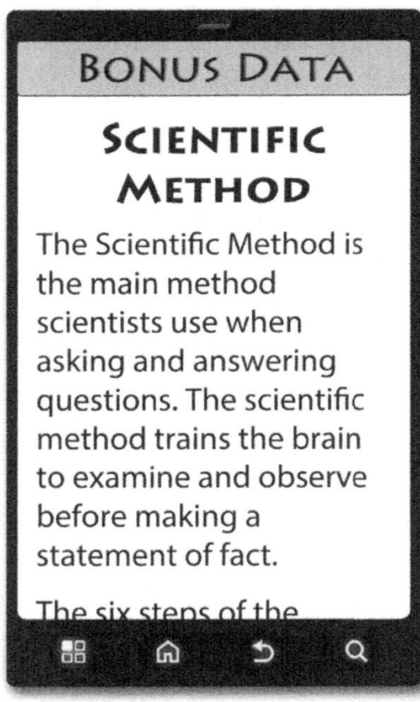

"We get bonus data this time, too, even though we're completely finished," Blaine mused.

"It seems like it," Tracey said with a smile.

"Yes, it does," Summer agreed. "Well, go ahead you two champions of science. Open up your backpacks, pull out those smartphones and check out that bonus data!"

Again, the twins exuberantly obliged the joyful scientists. Blaine was the first to see his screen and read, "The

Scientific Method. This is the main method scientists use when asking and answering questions. The scientific method trains the brain to examine and observe before making a statement of fact."

> **BONUS DATA**
>
> **SCIENTIFIC METHOD**
>
> The six steps of the scientific method are: ask a question, do some research, formulate a hypothesis, test with experimentation, record and analyze observations and results, and finally draw a conclusion.

Now Tracey jumped in. "The six steps of the scientific method are: ask a question, do some research, formulate a hypothesis, test with experimentation, record and analyze observations and results, and finally draw a conclusion."

"Hey, this reminds me a little of the engineering design process that Jaya Tang taught us about in Cambodia," Blaine inputted.

"Yeah, me too," Tracey agreed, and then read more. "Step one is to ask a question. The scientific method begins with asking a question about something that has been observed. Good questions must be about something that can be measured. These questions typically begin with how, what, when, who, which, why, or where. Step two is to do some research. One needs to read about the topic from the question so that he or she will have some background knowledge. This will keep the scientist from repeating mistakes that have been made in the past."

"Step three," Tracey continued, "is to formulate a hypothesis. A hypothesis is an educated guess about the answer to the question. The hypothesis must be measurable and answer the original question that was asked."

Tracey paused her reading and looked at her brother. "Okay,

your turn bro. Finish it off."

Blaine immediately obliged. "Step four is to test with experimentation. The experiment tests whether the hypothesis is true or false. It is important for the test to be fair, meaning the scientist may need to run multiple tests. He or she will change only one factor at a time in the experiment so it can be determined which factor is causing the difference. Step five is to record and analyze observations and results. During the experiment, the scientist will record all his or her observations. Once the experiment is complete, he or she will collect and measure all the data to see if the hypothesis is true or false. The scientist will often find that the hypothesis was false. If this is the case, he or she will formulate a new hypothesis and begin the process again until there is an answer to the question."

"Step six is to draw a conclusion. Once the scientist has analyzed the results, he or she can make a statement about what was found. This statement communicates the results of the experiment to others."

Upon finishing the bonus data, Blaine had drawn a conclusion himself, and he shared it right now. "I say it's time to party."

"I agree!" Summer exclaimed with a smile. "And evidently so do Socrates and Aristotle. Look! They're both wearing party hats."

Both twins looked over toward the resident mannequins and sure enough, both were wearing paper party hats, and both had sticky notes attached to their left hands with some kind of message written on them. The siblings rushed over to see what the pieces of paper said.

Tracey got to Socrates first and read what was written on the philosopher's sticky note. "Welcome back, Train and Blaisey! Sweet sassy molassey, you two have done it! Zip, zip, zooray! You've completed your summer of face-to-face science learning!"

Then Blaine read what was attached to Aristotle's hand.

"Pardon my absence, you two fantabulous zip-meisters, but I have gone to confess my undying love for Summer Thyme Beach at the reunion in the junior high school building!"

He was so excited he could hardly contain it. No. More. Days. Why? Because today was the day—the day he had been waiting for since he'd lost his eyebrows so many years ago in that junior high classroom. Today was the day of vengeance.

All of his checks this morning had shown that the Forget-O-Nator machine worked perfectly. He had set a festively decorated sign outside the door, complete with balloons, streamers, arrows, and exclamation points. The sign said, "Welcome esteemed guests to your junior high class reunion!" Inside the door, he had set up the Forget-O-Nator to look like a classroom, putting out some snacks, a punchbowl, and a few more decorations to make it look at least a little like a reunion might look.

Currently, he was standing in his Dark Cape suit. He would greet the invitees invisibly. The hope was that they would enter on their own and then he would slide in behind them, the door automatically locking behind them all. He would pull the vanish string, shocking them with his sudden appearance. He would inform them of why they had been invited and then he would wipe everyone's memories away.

The date, time, and location on the invitations had been clear. They should all be arriving soon. And when they did, vengeance would finally be served.

The Sassafras twins were so excited they could hardly contain it! Their uncle was going to confess his love to Summer! Finally! They'd been waiting for this to happen for a while, but of course Summer had been waiting even longer.

Upon reading the sticky note, Summer Beach had wrapped them and Ulysses up in a joyous happy jumping dance hug. Then the four had virtually flown up the basement stairs, through the house, and out the front door. In the process, they had bumped into President Lincoln and REESE. Now the group of three humans, two animals, and one robot were racing 1104 North Pecan Street to the junior high school building.

"He must've forgotten that I was your last local expert, and that I would be zipping back with you two," Summer laughed as she ran. "In his wonderful, lovey-dovey state of mind, he didn't realize if he would've just stayed put in his basement, we could've seen each other sooner. Oh, Blaine and Tracey, I have been waiting for this so long! I have loved your uncle since the first time I met him! And now I know he loves me!"

The twins were also smiling and laughing as they ran. They passed Mrs. Pascapali's house, sitting in her front porch rocking chair.

"Where are you youngins running off to?" The kindly widow asked.

"Cecil is going to confess his undying love for Summer!" Blaine shouted for the group.

"Well, isn't that nice," Mrs Pascapali responded.

On the opposite side of the street, Old Man Grusher also sat in his front porch rocker. He was accompanied by his miniature poodle, but today there were also three smiling young adults with him. Two males and one female. One male with a mohawk, the other with dreadlocks, and the female sporting multiple earrings.

"Hi, Mr. Grusher!" Tracey waved as they quickly passed by.

"Do you have some visitors today?"

The old man nodded happily. "My niece and her friends are paying me a visit to tell me all about their world travels this summer."

"Super cool!" Tracey replied as her group of six continued down the street.

His airplane had landed as scheduled. He'd rented a car and now Yang Bo was making his way from the airport to the neighborhood. He could've called one of his old friends to pick him up, but instead he wanted to surprise them.

The group of six sped past the Left-Handed Turtle supermarket, the Right-Handed Pelican Pizza Parlor, and the Ambidextrous Octopus Bowling Alley, and were about to pass the library as well, when they ran into Mrs. Womberfraggle.

"My goodness, aren't the six of you going somewhere awful fast," The librarian smiled and laughed.

"We are headed to the school for the junior high class reunion!" the twins said cheerfully.

"Oh yes, that's right," Mrs. Womberfraggle replied. "I remember hearing about that. Well, I am headed over to the school too, to meet my husband. Do you mind if I walk with you? Granted, I can't go as fast as you all were going."

"We would love it if you would join us, Mrs. Womberfraggle!" Summer answered for the group, here cheer level even higher than the twins.

Using top notch Swiss surveillance and Swiss army knife technology invented by Q-Tip himself, the two agents had located Paul Sims. The criminal curator appeared to be on his way to a junior high school building.

"All right Q-Tip, my friend," Evan DeBlose said with his normal level of confidence. "Let's hide in these bushes until we see him go in. Then, we'll follow in after him."

Q-Tip nodded his slightly oversized head, and the two sunglasses-wearing, dark-suited agents stealthily hunkered down in a hedgerow, and just in time too. Because here came five people, two animals, and a robot.

"Hello, my Love!" Ernest Womberfraggle cheerfully greeted his wife, Wilma. "And hello to you as well, Summer and friends," The junior high science teacher then greeted the others who'd entered the building with his wife.

"Hello, Mr. Womberfraggle!" Summer returned the greeting. "Do you know where the junior high reunion is taking place?"

"Well, I saw a sign for it in that direction," Mr. Womberfraggle pointed. "I'll show you."

The group of eight made their way through the corridors of the teaching institution until they reached the sign in the hallway the teacher had mentioned. It was a three-foot A-frame chalkboard sign festively decorated with balloons, streamers, and bright colorful letters that said: "Welcome Esteemed Guests, to Your Junior High Class Reunion!" The arrows were clearly directing those who came to walk through the door adjacent to the sign.

Summer took a big happy deep breath. Would Cecil be inside? Blaine and Tracey took in anticipatory breaths too, as Summer pushed open the door. With Summer leading the way, all eight entered the classroom, the door clanking closed behind them.

Cecil was not here. However, the place was set up for a party. There were decorations, snacks, and even a big bowl full of punch.

He quietly and invisibly snickered to himself. His first captives had been nabbed. The Womberfraggles, the rodents, the robot, Summer Beach, and—Blaine and Tracey Sassafras! Yes! He finally had those twins! He'd been trying to capture them all summer, and finally, he had them! They'd walked right into the Forget-O-Nator with no need for coercion. Yes, those twins, plus six, were now locked inside the Forget-O-Nator, but he would remain in the hallway, expectantly waiting for a red-headed guest to arrive. Oh oh! Here he came now!

The Reunion

The Womberfraggles sat down in some of the available seats, happy to wait for any others to arrive. Lincoln, Ulysses, and REESE huddled up together, also content to wait. Blaine seemed very interested in the punch bowl. Tracey, however, was a little concerned. She hadn't liked the way the door had clanked shut behind them, and now that she gave it a little pull, she found that it was locked. That didn't seem like a good thing.

Summer of course wasn't thinking about the punch bowl or

the locked door. She also wasn't excited about anymore waiting. She wanted to find Cecil.

"Where's Cecil?" Summer asked, eyes wide open.

All at once, there was a bump and a thump. The door opened again. The eight in the classroom turned to see who it was. They saw one bunny slipper, then two. A messy white lab coat. Red unkempt hair. The biggest smile a face had ever held. It was none other than Cecil Sassafras!

Clank. The door shut behind him. The uncle made quick eye contact with his niece and nephew. Even though it was only a split second, he clearly communicated how proud he was of the two for their immense accomplishment.

The scientist's eyes then locked onto the eyes of the other scientist. Summer's face now competed with Cecil's for the biggest smile ever. The witnesses in the room could almost hear the hearts of the two beating faster and faster.

The two seemed to float in the air, as if they were experiencing an emotion stronger than gravity. Cecil opened his mouth to make an announcement. Summer clasped her hands close to her heart as if she'd been expecting what he was about to say. It was clear to all that they were about to witness a long overdue confession of love.

His patience had run out, but not because of frustration or discouragement. His patience had run out because of excitement and coming victory. His plan was working! After all this time, he almost couldn't believe it! He already had the most important subjects inside the Forget-O-Nator. Yes, Paul Sims might eventually show up, but he had nine locked inside!

He could wait no longer. He would now step inside to greet them.

Creak. The door opened once more, the sound of it interrupting the confession between the lovebirds.

Clank. The door shut again without a new arrival. The twins shrugged and looked at each other.

Suddenly, the Dark Cape appeared! The villain towered over them with outstretched arms. The suit was midnight black from head to toe. Even its eye-protecting visor was as dark as dark could be. The gloves, boots, and body of the suit were also black, and the suit was designed to puff the wearer up and make them look bigger than they actually were. And the cape—the huge flowing cape whirled around the miscreant like it was being blown on by some unseen wind.

The Dark Cape balled his gloved hands into fists, raised

his arms even higher, and then let out a deep, bellowing, sinister laugh that could be felt in the soles of their shoes. Though they had run into him several times this summer, something about this encounter felt extra frightening to Blaine and Tracey. It was like the villain was twice as confident of himself as he ever had been before.

The twelve-year-olds fell to their seats. The Wombergraggles remained on theirs. The rodents and robot looked frozen stiff. Cecil and Summer remained standing, each facing the villain with looks of wonder more than anything else.

The wicked laugh subsided, and the Dark Cape lowered his hands to each side of his helmet. He gave the pointed aerodynamic helmet a click and a twist, and it came off.

"Thad Wazeezy?" Summer exclaimed.

"What was easy?" Cecil asked curiously.

"No," Summer said, pointing at the revealed eyebrow-less face. "It's Thaddeus Wazeezy, our old junior high classmate."

"It surely twirly is," Cecil now recognized the old, familiar face.

"The Man With No Eyebrows!" Blaine and Tracey wheezed from tight throats.

"The Man With No Eyebrows," Cecil repeated and paused with a look of revelation. "Thad. Thaddeus. Wait. This means you're the one—" Cecil's brilliant brain was making all kinds of connections "—who successfully copied the invisible zip line concept. The one who spied on us from that basement at 1108. The one who stole this magic suit. Are you the one who's been trying to stop Train and Blaisey from learning science? Are you the one who attempted to wipe away the memory of Sweet Summer—"

"Yes!" Thaddeus cried out in a bold voice. "It was all me! And I'm about to do it again."

"Again?" Summer asked.

"I'm going to wipe it all away. Everything!"

"Our memories?" The female scientist tried to clarify, now looking a little nervous.

"Correct," Thaddeus seethed. Then he started a more in-depth explanation. "Yes, you were all invited to a junior high class reunion, and this room looks decorated for such an event, but that's not what this is. That's not where you are. This, my esteemed guests, is actually an extra-large Forget-O-Nator. A bigger, better, more powerful version than the one I used to wipe away your memory, Summer, and almost your memory, Blaine."

Blaine made a face and pointed a silent finger at himself.

"But Blaine had a taser, and Summer had a back-up canister of stored memory, so the first attempt to wipe away your memories didn't work," Wazeezy snarled. "But this machine, this machine we are all trapped in now, is sure to do the trick."

"We're trapped?" Tracey got a desperate question in.

"Indeed," the eyebrow-less man answered. He was standing next to the door, and he now pointed at it. "The door to the machine is designed to automatically lock when it closes."

He then opened a panel on the wall that no one had noticed up to this point, revealing a large knob. "I will soon crank this one-and-only knob up to full capacity. The lights and power of the machine will surge, and in just one second, we will all be forget-o-nated. All the memories all of us have ever made will be wiped away! Everything we've learned will be gone. Every time we've loved, laughed, lost, or lamented, will disappear forever! All the victories, all the defeats, everything will be gone!"

"Wait," Cecil said, still very curious but somehow unalarmed. "So does the memory-wiping second happen during the power surge or after?"

"What?" Thaddeus asked. "That's so not the point, Cecil!"

"Welly-well, I was just wondering because I'm thinking through the scientific method."

"No, Cecil! There's nothing for you to figure out here. There's nothing to calculate or invent or experiment with! You've lost. I've won!"

"Wait," now Summer Beach chimed in. "You're in here with us, Thad. Are you saying you're going to wipe away your own memory, too?"

"Yes!"

This response from Thaddeus made the whole room pause. It was Cecil who then asked the next obvious question. "But why? Why are you doing this, Thad?"

"Why?" Wazeezy asked with a scowl and roguish curl of his lips. "I'm so glad you asked, Cecil. The answer: vengeance!"

"Vengeance? Vengeance for what?"

"Vengeance for taking my eyebrows!"

Everyone in the room was speechless. The Womberfraggles looked frozen. REESE looked robotic. President Lincoln and Ulysses S. Grant looked shocked. Blaine and Tracey looked confused and scared. Summer looked sad. Uncle Cecil still looked like he was trying to figure out what Thaddeus was talking about.

"You don't remember?" The Man With No Eyebrows asked flippantly. "Well, let me refresh your memory. That is, before I erase it!"

The unhinged villain released another wicked laugh before he continued. "It all happened one fateful day in Mr. Womberfraggle's chemistry cla—"

Suddenly the door creaked open again, and in entered . . . Paul Sims.

"Oh there you are, Summer! I thought we were suppo—"

"Paul!" Summer shouted. "Don't let the door shut behi—"

Clank. Paul looked at Summer, confused.

Then, he looked at everyone else in the room: The Womberfraggles, the Sassafras twins, Cecil, the animal lab assistants, the robot sidekick, and . . . was that? That face looked familiar. Was that really . . .

"Thad Wazeezy," Summer said.

"What was easy?" Paul asked.

"No, Paul. It's Thaddeus Wazeezy, our old junior high classmate. He's locked us inside this classroom, which is actually an enormous machine called a Forget-O-Nator. He is going to erase all our memories."

"Why?" Paul asked, more than a little confused and looking angry.

"Because he said Cecil took his eyebrows," Summer explained.

The museum curator now gulped and glared in Thad's direction.

"It all happened one fateful day," The Man With No Eyebrows began his monologue again. "In Mr. Womberfraggle's chemistry cla—"

Creak.

"Don't let the door—"

Clank.

Blaine had been the one to shout out about the door, but it had not mattered. The two who'd slipped in had let it shut.

The black-clad secret service agents removed their sunglasses and took a quick look around the room. They kindly waved to the Womberfraggles. They acknowledged the Sassafras twins, who were looking at them with wide eyes of recognition. They nodded their heads toward Summer, Cecil, and the three sidekicks with familiarity. However, when their eyes got to Thaddeus, Evan and Q-Tip looked utterly confused.

"We're not even going to ask," Evan DeBlose said, turning from the eyebrow-less man toward the museum curator. "Paul Sims, we're here for you."

Paul dropped his eyes and gulped again. "Why?" Paul asked weakly.

"Because of the crimes you've committed on earth and in space."

"Earth and space?" Sims chuckled nervously. "What are you talking about? I'm friends with Cecil Sassafras. You don't think a friend of Cecil's would commit any kind of crime, do you?"

The faces of both Evan and Q-Tip remained unreadable, with no answer.

"Additionally . . . I'm . . . I'm . . ." Paul dripped with sweat. "Dating Summer Beach! How could someone romantically involved with Summer, who is so deeply connected to Triple S, ever commit crimes on earth or in space?"

There were confused looks all around the room, but no one looked more confused than Summer.

"Enough!" Thaddeus interrupted, desiring to regain control of the room. "None of this even matters!"

"Sure, it does," Evan DeBlose begged to differ.

"It does not," Wazeezy bellowed. "I'm about to wipe it all away."

"I'm not following you, buddy," Evan said—calm, cool, and collected.

"You don't need to follow me," the Man With No Eyebrows snarled. "You just need to let me finish my story."

Thaddeus started again, "It all started one fateful day—"

Creak.

"Yang Bo!" Summer shouted happily.

Clank.

"Surprise!" The Chinese astronaut exclaimed, full of joy himself, not knowing the situation. "So good to see everyone. Wow, I only expected to see Cecil, Blaine, Tracey, and maybe Summer, but Paul! You're here! Long time no see, and Mr. And Mrs. Womberfraggle. And, hey! Is that Thaddeus Wazeezy? Hey Thad! What are you wearing, man?"

Yang Bo then turned toward the Swiss Secret Service agents and was about to introduce himself, but he was interrupted by The Man With No Eyebrows.

"So glad you could make it, Yang Bo. Now be quiet!"

The astronaut was more than a little taken aback by his old classmate's rude outburst.

"Thaddeus is about to wipe all our memories away," Summer said with a grieving face.

Yang Bo obviously didn't understand, but instead of letting any more explanation happen, Thaddeus Wazeezy once more jumped into his monologue. "It all happened one fateful day. In Mr. Womberfraggle's chemistry class." Thad paused for a second at this spot, halfway expecting someone else to walk through the door. But no one did. He smiled wickedly and continued. "We were in the lab, and it was the first time Mr. Womberfraggle had let our class use the Bunsen burners."

The teacher dropped his head as if he knew what was coming next.

"I was at a table by myself," Thad continued. "Because my normal lab partner was out sick. On my left were Paul Sims and Yang Bo. On my right were Summer Beach and Ccccecil Ssssassafrasss," Wazeezy hissed the scientist's name as he shared the story.

"I accidentally knocked my pencil off the table," he continued. "I bent over to look for it, picking it up. Then, as I stood back up with my face precisely table level, a flash of heat and fire blasted out at me! Both of my beloved eyebrows immediately burned off. The entire class laughed at me as I stood there alone, the sole victim of the explosion. Then, to make matters worse, Cecil tackled me and stomped on me."

"From that day forward," Thaddeus continued. "Cecil Sassafras became my enemy. I promised myself I would one day exact revenge on him. While I was subjected to a life without eyebrows, Cecil went on to become a pseudo-famous research scientist. Over the years, my rage grew stronger. My need for revenge became all consuming!"

Thaddeus shouted and shook his fists. "Cecil and President Lincoln eventually invented a means to travel the globe at the speed of light. He was going to use these invisible zip-lines to help his niece and nephew travel around the world to study science."

Blaine and Tracey looked around the room. The existence of the invisible zip lines was supposed to remain a secret. However, they were pretty sure that everyone here already knew about them, except for maybe Mr. And Mrs . . . hey, what was that in Q-Tip's hand?

"Taser!" the small agent suddenly shouted, lunging for the locked door of the Forget-O-Nator. Thaddeus made no move to stop him. Q-Tip stuck the taser of his pocketknife on the door handle and tazed away.

Some sparks flew, but almost as soon as they'd started, the sparks fizzled out. The only thing that remained was a tiny black spot on the door handle.

Thaddeus looked at Q-tip, and then the eyebrow-less man spoke, "That worked on the old Forget-O-Nator, but it will not work here. My host of scientists who designed and built this thing made it virtually indestructible. It is taser proof. It jams cell phone signals from inside, and it is inescapable. You are all stuck in here, and you are all going to have your memories erased. There is nothing that can stop that now."

Q-Tip shuffled back over next to Evan and hung his enormous head.

"Where was I?" Wazeezy wanted to get back to his long-winded monologue. "Ah yes, the invisible zip-lines. That's when I decided the best way to pour out a full measure of revenge on Cecil Sassafras would be to stop his niece and nephew from learning science!"

It all made sense now. Now they knew why this man with no eyebrows had been pestering them all summer. But he hadn't succeeded until now. Was he really going to wipe all their memories away? Was all they'd learned really about to be gone? They'd gone from despising science to loving it. Now they were just going to forget it?

"Vengeance is mine!" The Man With No Eyebrows shouted. "I will finally receive all my desired revenge!"

Thaddeus reached toward the open pane. "Now, I will crank this knob to full capacity. The lights and power of the machine will surge and in just on second we will all—"

"Hold up, Thaddeus. Let's back up a bit," Yang Bo said. "The way you described the incident in the classroom was close to what happened, but at the same time, it was way off."

Thaddeus was tired of getting interrupted, and he was about to put his foot down, but Yang Bo wouldn't let him.

"Yes, Cecil and Summer were on your right, but it wasn't Cecil who caused your Bunsen burner to explode. I can remember how it all happened, as clear as day."

"You can?" Tracey heard herself asked.

Yang Bo nodded. "I sure can. It was Paul Sims."

Everyone in the room gasped, even Thaddeus. Paul gasped at the accusation. Then he gulped, then he glared at his old classmate.

"Sorry to throw you under the bus, Paul," Yang Bo said to his old lab partner. "But I can't just stand here and let Cecil get falsely accused."

"What? How?" Thaddeus began stumbling through the forest of questions of a possible significant change.

"Paul planned the whole thing out," Yang Bo continued, sharing the true version of what happened that day. "Before class, he turned your Bunsen burner on, Thaddeus. Then he disconnected the hose, creating a gas leak. When you bent over to pick up your pencil, Paul tossed a lit match toward your table. I don't think he meant for the explosion to be as big as it was, but it was all intentional."

"But, but . . . but . . . why did Cecil laugh at me and tackle me and stomp on me?" Thaddeus asked in mumbles.

"I think Cecil was the only one who wasn't laughing. Summer and I weren't either. But Cecil tackled you and stomped on you because you were on fire, Thaddeus. It wasn't just your eyebrows. Your clothes were on fire and the top of your head too. You may have lost your eyebrows, man, but Cecil saved the rest of your hair, and he very well may have saved your life. You shouldn't be trying to exact revenge on Cecil. You should thank him."

Upon this revelation, Thaddeus Wazeezy's legs gave out. He fell into his seat completely and utterly stunned. The eyebrowless man seemed overwhelmed and relieved all at the same time. Everyone had gathered around him, with Cecil, Summer, and Yang Bo sitting right next to him.

"I . . . I've . . . wasted my life," he said looking down. "I've spent all this time and energy being angry at you, Cecil, and plotting revenge against you. And it turns out, you're the one who saved me. Yang Bo, you're right. I should say thank you."

The Man With No Eyebrows covered his face with his hands, rubbing at the tension knotted in his forehead. The twins watched as he wrestled with the recent revelations. His hands dropped and his spine straightened. Thaddeus looked up at their uncle. "Thank you, Cecil. Thank you for helping me that day."

"All's forgiven, Thad," Cecil patted his old classmate on the back.

"I know it would probably make sense for my anger to now shift to Paul, but—" Thaddeus continued to speak with more clarity "—I think maybe I'm just too tired. Too tired to be angry anymore. Too tired to carry around the weight of revenge or vengeance. I think instead, I'll just forgive him, too. Where are you, Paul? I just want to say I forgi—"

Thaddeus's sentence dropped off a cliff of shock. Paul Sims was not in the group surrounding the man. Paul Sims was standing over by the open panel on the wall with his hand on the one-and-only knob of the Forget-O-Nator.

"I want to forget it all!" Paul shouted as though Wazeezy's former insanity had been transferred to him.

"No, Paul!" The Sassafras twins shouted. "Don't do it!"

But it was too late. Paul Sims had cranked the knob to full capacity. The lights in the room flickered. Now Cecil's earlier question of when exactly the memory-wiping second happened was of utmost importance.

Ernest Womberfraggle tightly embraced the love of his life. He and Wilma had created a lifetime of memories together, memories that were about to be gone.

Evan patted his fellow agent on the back, "It's been a good run, pal."

"Sure has," Q-Tip agreed, knowing that the good run didn't matter anymore.

President Lincoln, Ulysses S. Grant, and REESE the Robot gathered together in a tight circle. If this was the end, the three friends were glad they were together. Yang Bo and Thaddeus looked at each other, and all they could do was laugh. The truth had finally been told, and the truth had set free. That was a good thing, even if enjoying the truth was going to be short-lived.

Blaine looked at his sister. Tracey looked right back at her brother. She was trying really hard not to cry. "Well, Trace, I sure wish we wouldn't have left our carabiners and harnesses in Uncle Cecil's basement. Now we can't zip out of here."

"You know we wouldn't zip away even if we had that stuff. We couldn't leave everyone else behind." Tracey said.

Then, instead of saying something funny or sarcastic, Blaine wrapped his twin sister in a hug. "It's been a great summer, sis."

"It sure has, bro."

Tears now came into both of their eyes. An angry hum sounded as the power of the machine surged.

Cecil and Summer faced each other, clasping each other's hands, and stared into each other's eyes.

"Summer Thyme Beach."

"Yes, Cecil? Yes?"

"I have an equation for you: $CS + SB \rightarrow L_2F_2$"

Summer's eyes sparkled as she took a delighted guess as to what the wonderful equation might mean.

Cecil then took the guessing out of it and said clearly, "Cecil Sassafras plus Summer Beach yields lifelong love and forever friendship. I love you, Summer. And I always have."

"Oh, I love you too, Cecil! I love you! And I always have!"

The long, overdue confession was followed by a happy jumping dance hug—the last happy jumping dance hug that would ever happen. The hugs, love, friendships, camaraderie, truth, forgiveness, and memories were all about to be wiped away forever.

The lights of the Forget-O-Nator continued to flicker. The power surged stronger. The memory wiping would happen in a flash . . . *creak*.

The door had creaked open and stayed open. The sound of a multitude of footsteps could be heard, but not seen. The one-and-only knob cranked back down, seemingly by itself. Paul Sims was knocked down, fighting as if he was being held down by an invisible force. The lights in the room returned to normal. The power stopped surging.

Everyone looked around. They knew each other. All their memories were still intact. What had just happened?

Suddenly, a dark cape appeared in the room. Then another. And another. Until there were scores of the caped figures standing in the room. Blaine's and Tracey's mouths hung open along with everyone else's.

"There's more than one Dark Cape?" Blaine muttered. "I thought the Man With No Eyebrows stole the only one from Phil Earp?"

Thaddeus Wazeezy's mouth hung open, too. "I knew about the extra dark capes. I just didn't know where they had all gone."

The eyebrow-less man looked around the room at all these figures suited up like him. "In the same lab this Forget-O-Nator was created, an army of Dark Capes was also created," Wazeezy explained. "But one day, the lab was inexplicably destroyed. The only thing that survived was myself and the Forget-O-Nator. All those Dark Capes used their specially designed three-ringed carabiners and the invisible zip lines to escape the destruction. Inside the suits, at least at that time, were scientists who worked for me. I don't know who might be in them now."

As an answer, one of the Dark Capes reached up and grabbed

his helmet. After a click and a twist, his face was revealed.

"Phil Earp!" The Sassafras twins exclaimed but then repeated it again in question form. "Phil Earp?"

The magician, who the twins had met in Sydney, Australia while studying anatomy smiled. "The one and only," he chuckled.

It was a match made in heaven. It was easy to see that he and she belonged together. They were like bacon and eggs. Like a tire swing and a tree. Like a novel and an easy chair. These two were made to go together.

With enraptured smiles on their faces, the swooning couple walked hand in hand down the quaint neighborhood sidewalk. The afternoon breeze and sprawling rays of sunshine accompanied them as they took unrushed steps, fulfilled and content in each other's company. Their gently clasped hands slightly swung back and forth as they walked, matching the rhythm of their smiles. Cecil Sassafras and Summer Beach were finally together, as they should be.

Blaine and Tracey walked behind the giggling and whispering love birds. "They're in love, Blaine. They know it. We know it, and now everybody knows it," Tracey said happily.

"Hey Trace, do you think this means it's soon gonna be Aunt Summer?"

"Oh, I sure hope so!"

Everyone had been excited to leave the Forget-O-Nator with their memories, except Paul Sims. Mr. And Mrs. Womberfraggle had left the junior high building smiling and holding each other's hands. President Lincoln, Ulysses S. Grant, and REESE had scurried off beaming, happy to be out of the machine. Yang Bo had left to go grab a bite to eat at the Right-Handed Pelican Pizza Parlor with his old classmate, Thaddeus Wazeezy.

Triple S agents Evan DeBlose and Q-Tip had handcuffed and loaded up the only person in the room who had not been elated, Paul Sims, into a waiting heliquickter. Then they'd taken off to an undisclosed location where Paul would face punishment for his crimes.

Everyone had profusely thanked Phil Earp and the sizeable group of Dark Capes for rescuing them from the Forget-O-Nator. It turned out that everyone in a Dark Cape suit, except for Earp, had been a scientist in Thaddeus's underground lab. When they escaped the lab as it was being destroyed, they had all calibrated their specially designed three-ringed carabiners to the same longitude and latitude coordinates, which had zipped them all to Sydney, Australia. Phil Earp, who was in the backstage area of the Sydney Opera House, had been more than a little surprised when all the Dark Capes appeared right in front of him. Not only had a Dark Cape like the one that had been stolen from him been returned, but a cool hundred had shown up.

Phil Earp's current dream was to perform and star in a new magic show entitled 100 Capes. However, before that happened, the scientist knew they needed to stop Thaddeus and any nefarious plans he had. So they'd used satellites to locate him. Then, they had zipped together as a group and had shown up just in the nick of time. They were pleased to see The Man With No Eyebrow's drastic change of heart.

So with everything right in the world, all the Sassafrases, plus one Beach who might soon become a Sassafras, headed back to Cecil's house to celebrate a victorious summer of scientific learning. The four hopped the steps, sidestepped the trapdoor, and burst through the front door. The twins were exuberantly shocked by who they saw there in the living room.

"Mom and Dad!" Blaine and Tracey exclaimed happily at the sight of their parents.

EPILOGUE

Three Hours Later

Twelve-year-old twins Blaine and Tracey Sassafras, sat in joyous wonderment in the back seat of their parent's car. They were on their way back home after spending the entire summer studying science face to face with the help of their wonderful and wacky Uncle Cecil. They had used invisible ziplines to travel around the globe, and even into space, studying the subjects of zoology, anatomy, botany, earth science, geology, astronomy, chemistry, and physics. They had made friends all over the planet and had been helped and guided in their learning by extraordinary local experts.

They had celebrated the accomplishment thoroughly over the course of the last few days, and now they were headed home. Their parents were very proud of them. They had made that clear during all the celebration. They had also instructed Blaine and Tracey to write thank you notes to all the local experts when they got home.

Right now, their dad glanced at them in the rearview mirror and smiled. Their mom turned her head to look back at them, smiling also. "We are so proud of you two," she said. Blaine and Tracey beamed.

"That we are," their dad agreed. "However, we have something to talk to you two about."

Uh oh. Now the twins' faces dropped a little. They weren't sure what to make of the tone of their father's voice.

"We've come to a decision about your education. And you two are not going back to school," the twins' dad said.

Blaine's and Tracey's mouths dropped. "We're not?" they asked in unison

"We've decided to make a change. Instead of sending

you back to school, your father and I have decided to . . . start homeschooling you!" their mother said.

Blaine and Tracey looked at each other, exchanging joyful smiles.

Three Months Later

The old cowboy climbed the steps and knocked on the door. He waited in the whistling wind for a second. The door was answered by a fellow who looked just as Blaine and Tracey had described him. The cowboy tipped his hat and began his introduction.

"My mother named me Sylvester Robert Hibbel. My satisfied clients call me Doc. My wife calls me selfish or grumpy, but you may call me friend."

The man at the door responded with a slight chuckle to the unusual greeting.

The cowboy reached in his satchel and pulled out an old-timey glass bottle.

"This, my friend, is Old Doc Hibbel's Hair-Be-Here Sproutin' Sap. Just pour this solution on the desired area, and after an hour or two, hair be here, hair be there, hair be everywhere! This touting' sproutin' solution has been known to put hair on men and women in different stages of balding. It'll sprout hair on a shiny head. It'll sprout hair on a thinning head. Shoot, it'll even sprout hair on a watermelon if you want it to. And if I were a bettin' man, which occasionally is the case, then I would bet this will even sprout hair where eyebrows used to be."

Thaddeus reached up, rubbed his browless forehead, and smiled.

Have you missed a moment of the summer-long journey?

Don't miss a moment of the science-learning adventure! Follow Blaine's and Tracey's and zip-lining journey around the world from zoology through physics.

Visit ElementalScience.com/Sassafras to complete your Sassafras Stack.

Made in the USA
Monee, IL
13 September 2026

62980092R00193